ASTROPHYSICS
FOR
NON-MATHEMATICIANS

How to picture this universe at Bizarre spacetime curvatures

HITEN SHELAR

Made with ♥ on the Notion Press Platform

www.notionpress.com

Contents

*Wormholes (Page no. 79)

7) The Multiverse Hypothesis And The Many Worlds Interpretation.

8) The Participatory Universe

9) Are we alone in this universe?

ABOUT THE BOOK

Think Out Of The Dot !

Humanity was born on earth and is never intended to perish on it. Zooming out at the cosmic scales, we are nothing but just a dot-sized object. See the world around you, feel the running time by the looking in the past and ask the question where are we, who are we, and what is going on when we are somewhere which we don't have a complete sense of. This book attempts to answers these questions and also gives an idea how weird reality is .

The language of this book is oversimplified although the content requires high imagination power. The use of mathematics is avoided in this book and even if it is at some places, the reader has an option to skip it. Thus making this book a boon for beginners and enthusiasts who want to know the unimaginable secrets of our universe.

Preface

The real beauty of astrophysics is the way it seems tough to imagine or visualize the same universe we are residing in and that makes me laugh sometimes . And when we think mathematically, it just takes over a major part of imagination and delivers the results we needed. But there is no competition to the physical imagination. May not every time one gets the answer by physical imagination but that failure itself makes the study of our universe much more beautiful

A black hole is an infinite spacetime curvature, if is crossed then the dimensions of space and time seem to switch their roles! And one starts perceiving time with a freedom same as one has of various directions in space. What if I tell you that like positive and negative charges, mass also has its negative version? Imagine a particle with a mass having no positive, nonzero but a negative mass. These are one of the only mysteries of this universe that we know! and there is a lot much stuff out there to explore in fact we haven't even barely begun. 'Astrophysics For NON- MATHEMATICIANS' will take you far beyond our local scale. It will take you to places where the laws of physics seem useless and even very weird to imagine. At the end of the book, one will acquire a different vision of what reality is. This book combines my knowledge of astrophysics from various sources and the power of imagination. Remember, Humanity was born on

Earth and is never intended to perish on it! I will show you throughout this book how our mere existence as observers has shaped this universe into the way it is. After writing the last chapter of this book I hypothesized that every star, galaxy, and a black hole in this universe is made to make available fuel to the very advanced human beings in the far future (literally !).

. Throughout my book, I have used the physical approach much more than the mathematical one as I know how tasty the physical illustration is and also believe that the readers throughout the book would enjoy this recipe of understanding the fabric of spacetime.

Love is the one thing we're capable of perceiving that transcends dimensions of time and space

- Interstellar (2014)

Some Basics At Very Introductory level

What Is The Fabric of Spacetime?

As school students, we all may have done a bit of coordinate geometry. Basically in two-dimensional coordinate geometry we used to locate a point in space by two numbers and by three in three-dimensional geometry. Well, what about describing the real-life events by the same coordinates? Let's consider the event of the arrival of a train and the ideal the question we ask is where will the train arrive and at what time? To emphasize , what you are asking here is what is the location in space where the train will arrive and at what time. For the description of real-life events, one needs time to be treated as a separate dimension. Having three spatial coordinates (x, y, z) and one-time coordinate t, the events can be successfully described. Well, this space and time in reality, is spacetime and not a separate thing, they are connected to each other.

Spacetime (x, y, z, t) is a four-dimensional entity (three spatial and one time) and of course we humans can't see time as it's not spatial. One can be in rest in space, but time still be continuously moving on (exceptions to this will be later discussed). Spacetime can be thought of as a mathematical object to describe the events and its what our universe entirely made of. spacetime had its origin in the Bigbang (initial fireball). We cant see spacetime as it's purely just a mathematical object, but we know that we live in three dimensional (spatial) world and we can feel time as the dynamic evolution of this three-dimensional world around us. Spacetime geometry is not static, it can be made to change its shape by the curvature carrying effect of mass energy (later will be discussed in the general theory of relativity) .

The Universal Constraints

For the existence of inhabitants of this universe and in fact the universe itself, several universal constraints are made by mother nature. In reality, it is the fact that we as the inhabitants of this universe and the universe as our home have an influence on each other in a way that several rules and constraints are followed.

What Constraints?

There are many universal constraints in this universe to talk about (which is the part of another chapter), But what one will have in this chapter is the idea of the logic behind their formalism. Some of the examples of universal constraints are the speed of light constancy, shortest achievable length, mass-energy density constraint, etc

But why do we have these constraints at the first place?

Universe or to be more clear spacetime has a maximum limit of curvature it can exhibit. If the curvature goes too wild say until we get some mathematical infinities over there, then that part of the universe ceases to exist. It would be even rough to be called a part of the universe itself. Provided we have several universal constraints, not every system achieves the just mentioned mathematical infinities. The reason behind having universal constraints can be interpreted as the attempt of the universe to peacefully coexist with its inhabitants.

Why are scientists afraid of mathematical infinities?

Mathematics seems to be describing our universe so effectively to consider anything else. In fact, mathematics is the language of the universe. Physics as a subject has several rules to teach and mixing mathematics to it crystallizes it into the development of the world. Given the initial information about a soccer ball, one can describe its trajectory with the help of the laws of physics and math. The appearance of the mathematical infinities is not a fault of mathematics but could be due to our formalism of a particular theory to describe the system. If we manage to change or adjust our theory, mathematics would go fine over there.

'Strictly speaking, There is nothing special about the speed of light and it's the speed of CAUSALITY which is constant and not the speed of light!'

Speed of causality is basically the maximum speed anything could be able to travel at. The value of this constant is approximately 3,00,000 km/s. When one starts accelerating he or she tends towards 'c' and accordingly has its mass exceeded. Correspondingly that particular mass has some spacetime curvature associated with it which still changes when mass adds up to the system. Relativistic effects like mass, time dilation, length contraction, etc are basically the attempts of nature to make look speed of causality as same regardless of any frame. More clearly, The Fabric of spacetime changes itself in a way such that the speed of causality appears the same to every observer in this universe. And this induction of changes in spacetime is what appears to us as the relativistic effects.

Like the speed of causality, many universal constants play a role in maintaining this universe along with its inhabitants.

The Special Theory of Relativity (Introduction)

Let's start with the frames of reference. There are two frames of reference inertial (at rest) and noninertial (moving). Let's consider a train where Alice is enjoying his game of throwing a ball upwards and bob who is seeing his friend Alice from the train platform. For Alice, her ball is going upwards and for bob, his ball is tracing a curvilinear path as the train is moving (As Bob is in the rest frame). The action of throwing that ball looks so different in both the reference frames. One can also refer to Einstein's famous lightning bolt experiment for this. An event perceived by two observers is completely dependent on the spacetime coordinates they posse which is where you are and where you will be at a given time. Let's take another example where Alice and Bob are travelling in two individual cars at speeds v and u (v > u). Now according to Alice, Bob is travelling at speed v-u. But a person at rest on the footpath will still see bob travelling at v. So according to Alice, Bob is travelling at a slower speed than what it seems to be in inertial frame of reference. Now if an inertial observer flashes a beam of light travelling at C, what should be its speed with reference to Alice and Bob (think !). The answer is C (It's not a typing mistake). This is the law of constancy of the speed of light in vacuo (vacuum) .

Well its not speed of light in generality but the speed of causality (cause and effect) which is constant and has a value of *c* . Our Universe has constrained its participants with this speed limit. It will take an infinite amount of energy to reach this speed for us, which is practically not possible. But , What approach universe make to preserve its own set of rules or constraints made for its residents ? The moment Alice starts accelerating herself she starts getting towards the speed of causality, and as she accelerates more, the world around her (outside the train's window) starts changing itself so that the speed of causality or speed of light remains constant for her. As she achieves a speed closer to the speed of light she will notice shortening of the length of the objects outside her frame (Fitgerald contraction) . One more approach the universe makes to resist anything violating its constraints is the time dilation effect due to which after the journey has completed Alice and the pedestrian (inertial frame) won't agree to the total time passage anymore. More accurately, spacetime will dilate from its initial status of rest. As soon as one starts accelerating, her spacetime coordinates match no more to the one who was initially rest with her. These effects occur for a non-inertial frame as every inhabitant of this universe is constrained by the universe by the universal speed limit (the speed of light or causality) c . Special relativity has a set of transformations (equations) that connect inertial and non-inertial frames which calculate the relativistic effects like length contraction , time dilation , etc . Effects of special relativity are rare at the macroscopic scale and at the microscopic scale it's a normal thing to observe as it takes comparatively less energy for subatomic particles or particles with less mas to to achieve the speed near to the speed of light . The special theory of relativity is a vast field of study in Astrophysics and Quantum Mechanics.

The General Theory of Relativity (Introduction)

Newton in his time explained gravity as the line of force acting between two masses A and B . But in reality that is not quite true . His equations although work well in ordinary cases or say in some limit (Newtonian limit) , globally they fail . His equations doesn't explain what gravity is and where it got originated from. Einstein's famous General theory of relativity explains it greatly. According to general relativity, gravity is the curvature in spacetime. What brings this curvature is the energy or mass density. Einstein once had a thought experiment of an upgoing and free-falling lift from which he discovered that gravity is same as the acceleration which was the turning point of relativity. As I mentioned in the last section that when one starts accelerating, his spacetime coordinates start changing from its initial rest frame and if gravity is the same as acceleration! Can it be interpreted as a change in spacetime coordinates? Of course! It's like if one starts accelerating her spacetime coordinates will change and vice-versa i.e. if one's spacetime coordinates are changed he will start accelerating! That's great! Energy and mass density curves the spacetime (from the flat one) and spacetime curvature gets mass-energy into the acceleration . If Bob has his spacetime coordinates initially flat then as soon as he enters curved spacetime he will feel the acceleration (as space-time is curved), which is the gravity itself.

Well! Like a sheet of rubber, spacetime can be bent anyway one likes, all you need is the energy and mass and density at astronomical scales to curve it. Again the General theory of relativity like newtons theory of gravity is not universal like at the center of the black hole and what holds there is the Quantum Mechanics . Mass and energy density has an effect on spacetime by curving it and spacetime has an effect on mass by giving it a direction in its motion . We all are in the influence of the common spacetime curvature (by earth), our spacetime coordinates thus are identical. Everything in this universe traverses a geodesic in spacetime, which may be thought of as the shortest path one can have in spacetime flow. When spacetime is curved time dilates due to which time on Jupiter ticks slower than the earth (as Jupiter due to larger mass deviates spacetime more than the earth). Your flow of spacetime according to your frame of reference will never halt. This is the Principle of Relativity "Law of Physics Will Always Remain The Same In Any Given Frame". Whichever complex spacetime manifold you get into, you won't feel the relativistic effects of spacetime curvature unless you communicate with an inertial frame (outside the spaceship window) Thus it can be said that the speed of light and the laws of physics are not relative. General relativity math is all about calculating the spacetime curvature created by a given mass-energy density, by applying various coordinate transformations.

Black Holes

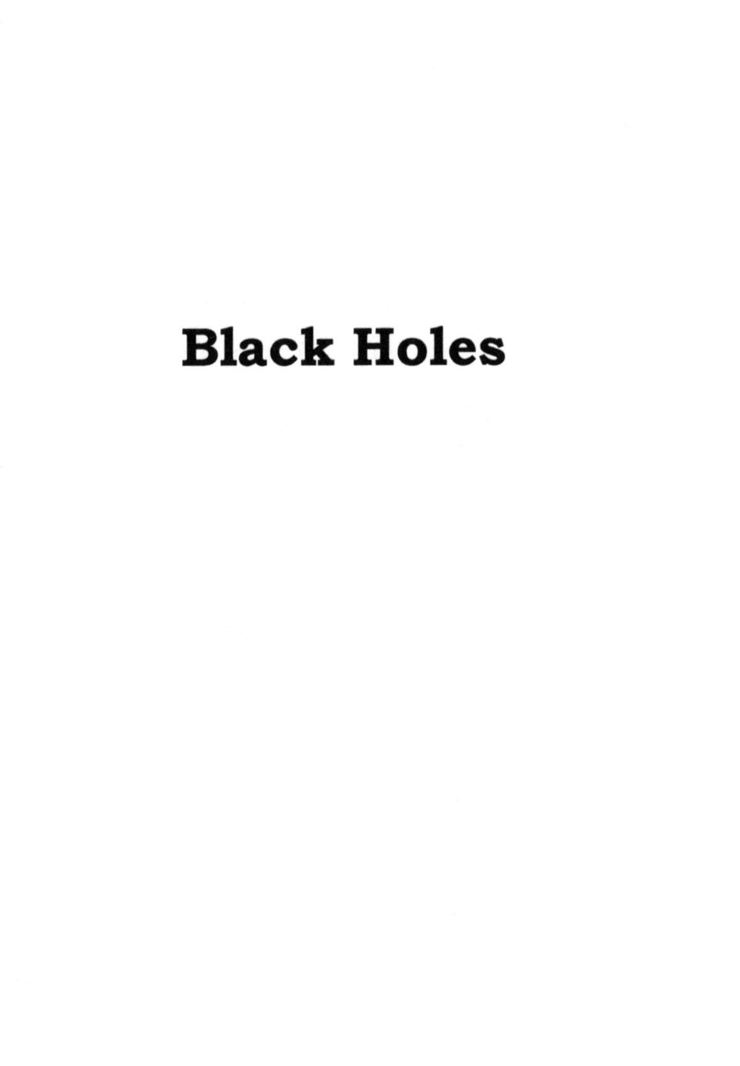

The Formation of Blackholes

Let's take a story of a star whose mass is say at least or more than 8 solar masses and is enough young to have the plenty of hydrogen to fuse. Remember at the scale of 8 solar masses the net gravitational force is very strong and we have a deep spacetime curvature associated with that star. For a star to maintain its surface from the strong gravitational force (inwards due to its own gravity) it needs some outward pressure to maintain the stellar surface . This pressure is carried out by the fusion energy in the star and the equilibrium produced between inwards gravitational collapse due to star's gravity and outward force is called the hydrostatic equilibrium. Due to the immense gravitational force inside the star , the atoms inside it move fast enough to start the fusion which is the source for external pressure maintaining the stellar surface. Starting with the fusion of hydrogen into the helium. Eventually, the time comes when that star runs out of hydrogen (as almost every hydrogen atom in the star fuses into helium), the helium produced inside a star is now used as a fuel and is converted into carbon after which carbon fuses with helium to form oxygen, oxygen plus helium produces neon, neon plus helium produces magnesium, magnesium plus helium gives silicon, silicon plus helium gives sulfur, sulfur plus helium gives argon, argon plus helium produces calcium, calcium plus helium gives titanium, titanium plus helium gives chromium and at last chromium plus helium produces the iron. Well! after that one cant really fuse iron into any other element, even after having such an extreme force (As iron is a very stable element). But still, if the inward gravitational force dominates over the incompressibility of iron then the iron atoms get enough close to each other until the star converts itself into a white dwarf.

A white dwarf

A white dwarf is a type of interstellar object which forms after the process as just discussed , Ir is what our sun is supposed to become, and it's a pretty stable object. The resultant pressure which prevents the white dwarf from the further collapse due to strong gravitational force is the electron degeneracy pressure. As soon as the star turns into the white dwarf and the inward gravity tends to collapse it still further, the degeneracy pressures attempt to maintain the white dwarf by giving a kind of a bump that shreds the outer layer of the stars into a violent supernova. Supernova would be a very beautiful event to look at in the sky if one can look at it at a safe distance. We humans have seen supernova with naked eyes in china on July 1054.

Is the electron degeneracy pressure result of any underlining fifth fundamental force? No. It's a quantum mechanical property for any two electrons or more to not to achieve an identical or the same quantum state, this is the Pauli exclusion principle. When the iron atoms in a star are forcibly being squeezed into each other, the individual electrons are here forced to share the same quantum state (violation of the Pauli exclusion principle), which is not possible. Thus to tackle with this the degeneracy pressure can be thought of as just the mathematical way or the tendency of our universe to avoid two electrons sharing the same quantum state. But what if the gravitational force still dominates? (Imagine the scale of spacetime curvature!) Now the only way for an electron to escape from this force and not to let the fundamental Pauli exclusion principle be violated, it has to combine with the proton which produces a neutron and a neutrino particle.

A Neutron star

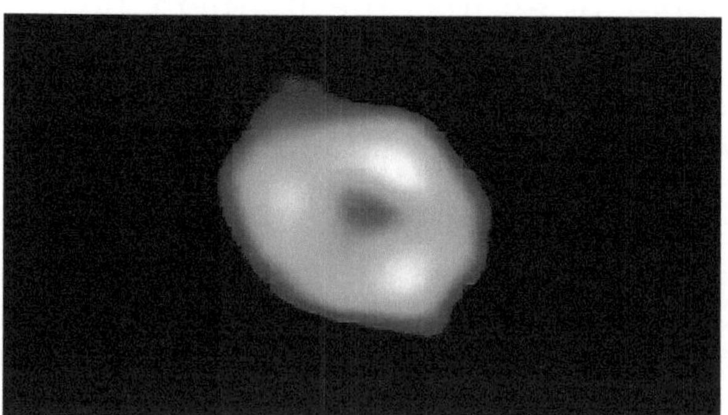

A black hole (Beast!). The image is the real image of black hole at the centre of our Galaxy.

Now we have a neutron star, which is almost entirely out of neutrons. Even neutron stars have their own degeneracy pressure, which maintains them from collapsing into a beast. Beast? yes! if gravitational force still dominates, then the star cuts itself off from the universe and we have a black hole there (A BEAST!). This is the recipe of the black hole, what you need here is just a star massive enough to cut itself from the neighboring universe .

Black holes and the Fabric of Reality

A point of no return from which even light can't escape. I am talking about Black Holes. Black holes can be thought of as an Infinite space-time curvature, where at the singularity General relativity fails! We know that light always traverses the straight path, so not does near the black hole as the fabric of reality (Space-Time) is itself curved there, making light traversing a curved path. Light basically is the oscillating Electric and Magnetic Fields in space or photons (packet of energy) that carry some energy and as spacetime is curved it has an effect on that energy flow. Other than this light always traces a straight path. Black holes are misconceptualized as two-dimensional holes. It's an infinite curvature in four-dimensional space-time, curvature in three-dimensional space makes it a spherical hole and a hole in time is not what we can see. The flow of spacetime inside the black hole can be compared to the water flowing into the drainage in the washbasin. What makes the black hole unique from other celestial objects in the universe? The thing is that for every celestial object we can successfully apply the field equations of Einstein to get spacetime curvature, at the center of a black hole one gets Bizzare solutions that are not acceptable. This point is famously known as the singularity.

What I think is, whenever the participant of the universe violates its laws it cuts itself from the universe, for example, we haven't detected any particle yet traveling at ultra-relativistic speed as they have already got themselves out of this universe. Similarly, what I think, is whenever an energy mass density makes up a particular spacetime curvature whose value seems to be violating the limit insisted by the universe it cuts itself from the rest of the universe and this is what we perceive as a black hole. We have all sizes of black holes in our universe ranging from microscopic to ultramassive black holes. Black holes are black and thus difficult to target, they are detected by the effect they have on their background like lensing of light due to spacetime curvature. Nothing is permanent in this universe and so does the black hole. Black holes do evaporate, they emit Hawking radiation which takes their mass out of them.

Let's take one interesting journey of an astronaut Cooper, who decides to get into the black hole to explore it. As cooper nears the event horizon time starts dilating for him until he stops at the event horizon. An inertial observer much far away from the black hole will be tricked as cooper never passed the event horizon, but as the principle of relativity holds cooper according to his own frame has passed the event horizon successfully. So who is real cooper inside the horizon or outside it? (think!) the answer is both the observers are correct (That's crazy!). The events one witnesses are completely relative, they are dependent on the space-time coordinate of one posse. Thus there is no reason here to agree with one of the observers as there is nothing as the benchmark event in this universe , If one gets her spacetime coordinates changed one can even dodge a particular event. Spacetime is the fabric of reality, changing it may result to change in the reality itself for that particular observer or frame of reference.

How Does the Black holes evaporate?

The beasts which swallow the fabric of reality itself. I am talking about the black holes, but they not only gain but also lose their mass. At the atomic or even smaller scales of spacetime, a pair of virtual particles pop into existence out of nothing, and again after a very small interval, they attract each other to annihilate themselves. Heisenberg's uncertainty principle allows their existence. It's like something out of nothing and again nothing. Does not it violate the first law of thermodynamics (energy can neither be created nor destroyed?) Yes! but finally No! Here, the fundamental first law of thermodynamics is just momentarily violated, and as a final result when particles annihilate each other, the energy is given back to the universe. Suppose we have an electron and a positron being popped out of nothing, creating negative energy for a limited time after which they annihilate each other, returning the energy to that negative energy density created so that finally the total energy created always remains zero.

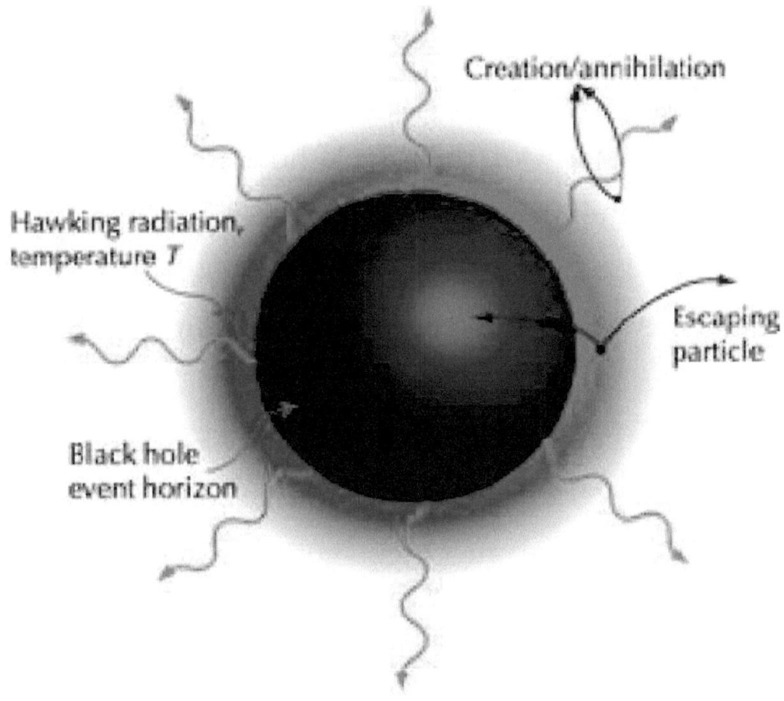

Analyze this condition around the black hole, as soon as the pair of virtual particles pop into existence they get separated from each other and one of them falls inside the black hole and the other gets away from it. This divergence in the middle of the path (where particles are approaching towards each other) is due to the tidal force of the gravity exerted by the black hole. The particle which gets away from the black hole gets its boost from the black hole itself! (as the tidal gravity results from the mass energy density of the black hole itself). And as mass is the same as the energy, that accounts for the loss of mass of the black holes. The particles which get thrown away from the black hole can be detected by us in form of radiation, this is the Hawking Radiation. We have never observed the Hawking radiation by our detectors. Once an object gets inside the black hole, its information completely gets lost, scientists speculate this information which is supposed to be forever lost inside the black hole to be carried out by Hawking Radiation. Even though black holes lose mass via Hawking radiation, it's still a very very long process for a black hole to get evaporated completely. For a supermassive black hole to get evaporated completely it would take 10^100 years (a very very very long time). As black holes eat a way more than what they lose by the Hawking radiation . Shortly , nothing is permanent in this universe and even the black holes perish by means of Hawking Radiation .

The new image of Sagittarius A * black hole.

On 12 May 2022, the first image of Sagittarius A* which is the center of our galaxy was taken, by the event horizon telescope. Event horizon telescope has done this previously when it took an image of an M87 black hole which is 2000 times larger than SgtA*, Scientist back then had their plan to capture SgtA*, but it's too far and there is a lot of interstellar material hindering the light rays coming from SgtA* to get detected as sharp as it would get M87 black hole. M87 is large and there's relatively there is less interstellar material to block the light. But now here in 2022 we have our dream fulfilled, the event horizon telescope has captured an image of SgtA*. It's blurrier than the M87 image which was taken in 2019 (due to the reason I just mentioned). To take an image of such a deep object in space we need a telescope having the size of the earth approximately, thus a telescope equivalent to that was designed by the connecting the radio observatories around the world. And now we have two products on our hands.

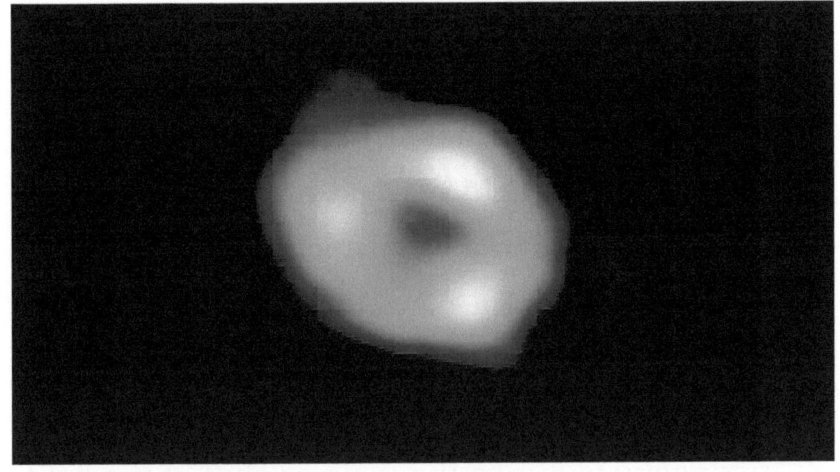

Sagittarius A* black hole

Description of the image :

Sagittarius A* has an accretion disk around it, which contains a pile of interstellar matter, gases, etc. material inside the accretion disk travel at relativistic speeds and has a high temperature associated with them, Blackhole is the region in space from which even light can't escape, but there is a point in the neighbourhood of black hole where light escapes the black hole by just a few minutes that region is located at 2.6 R (where R is radius of event horizon of the black hole). `Light emitted by the accretion disk which is behind 2,6R gets into the black hole and light emitted at and beyond this distance spirals in a way which, makes it appear as a ring around a black hole. One may notice that this ring is not continuous in terms of luminosity. This apparent fluctuation in luminosity is due to the doppler shift of the light. As the material in the accretion disk is traveling around at relativistic speeds, the material which is flowing say into the plane has its light red-shifted and otherwise blue shifted. The material in the accretion disk has been sucked by the black hole long ago, but it is still seen around the black hole in form of a ring around it.

Falling Into A Black Hole

Let's start our journey to what is inside the black hole with no time delay! Currently, we are in the accretion disk of the black hole paving our way inside it. As we get nearer to the event horizon time gets considerably dilated for us, which ultimately stops at the event horizon. Outside observer (inertial frame) will observe you as someone who never crossed the horizon, even if you in reality have had done that. a way before. As soon as we near the black hole one will see that before falling behind the horizon, some black background obscures our view of the universe around, Why is it that so? That's due to the curved path traced by the light near the horizon. To elaborate, light always travels a straight path in reality but provided that the spacetime metric is itself curved which governs the dynamic evolution of light itself makes the light curve. But the observer who shares that particular curvature of spacetime of light which made it get bent observed light to travel always at the straight path (These are relative concepts). As we near the horizon, the light rays which were traveling straight for some time before we neared the horizon, now take the curved path, and all the background universe be like projecting themselves in form of a particular circular patch in space dynamically eclipsed by the black background as we near the horizon. As we almost get over the event horizon the background projects itself as a point until it's all gone when we touch the Schwarzschild radius i.e the radius of the event horizon. This is a great view! when we were in an inertial frame we looked at the black hole as a small dark circle in the surrounding universe and as soon as now we were near the horizon it presents itself as a small circular universe in the surrounding darkness !.

Let's continue our way inside the horizon and now we don't care what the external observer sees! Spacetime switches their roles after one crosses the universe, here is where time is what can be perceived as space and space as a time. Time being perceived as space, one can see past, present, and future as we see the world around us. Well, unless and until we have a bulk being to make a tesseract for us inside the black hole. (as it was done in the interstellar movie) we may enjoy this strange world where time and space had their roles switched! Otherwise, we have to make our way to the singularity. Adding some points before we proceed on our journey to the singularity. The total charge, angular momentum, and mass-energy in this universe remain constant and that is the very fundamental law of this universe. Seemingly every particle which goes inside the black hole has its information lost. But to keep the fundamental laws of our universe unviolated, black holes keep all the quantities of angular momentum, mass energy, and charge surmounted over it. To elucidate, let's take a particle with charge Q, mass M, and angular momentum L. let's have this particle inside the black hole and the black hole gains the same mass as the particle, the same charge, and even the angular momentum gets added! That's so cool! Thus having somebody with angular acceleration with it when falls into the black hole, then according to the conservation of angular momentum, black holes will add this angular acceleration to itself by rotating faster just to keep conserving the law of conservation .

Falling Into A Black Hole-2

Before continuing our journey to the singularity of the black hole, let's understand what a black hole exactly is. The problem we have in describing the black hole is its singularity part. It's the infinite spacetime curvature and we don't know what infinity is! according to Quantum mechanics, nothing can be smaller than the Planck length. Classical and general relativity at this Planck distance fails to describe what is going out there. In fact, singularity is the point where spacetime loses its meaning. Wait! The universe is spacetime and no universe means what? It's unimaginable! It's like the pre-bigbang condition where time and space had no meaning. Let's still continue further, on our journey to the singularity and as soon as we get there our spacetime coordinates lost their meaning, but still according to the principle of relativity would laws of physics still hold in our own local frame? If No then it's unimaginable to think beyond this! If yes, then there is something for us to experience. Something but what? we have to know the evolution of spacetime after that and as there is no concept of spacetime at the singularity, we can't predict the physics beyond. Thus we may have to physically go there and check that. If Planck length dimension is the dimension which singularity has, then say at scale which is 100 times larger than this one has the quantum foam. This is the region or scale of the spacetime where everything runs in terms of a probability for example a universe may puff out into existence and simultaneously close itself and many more. Its also speculated that our universe started with a random fluctuation over this quantum foam which again collapsed into the supermassive black hole singularity and which in fact we call the cosmological singularity from which our universe began. Thus if we reach singularity then we may encounter the quantum foam which looks similar to like this:

The quantum foam (Strictly speaking, It may be of any shape!)

Remember! The dynamicity in quantum foam is time-independent (Think!). Throughout our journey, I have tried to explain what is inside the black hole by using the general theory of relativity and quantum mechanics, which in fact have their own contradictions.

Throughout our journey, the knowledge I served were the mathematical predictions. Mathematics is a tool that has a range throughout the universe without getting that particular part physically explored. A real-life voyage to what is inside the black hole is not suitable as the tidal force of gravity may

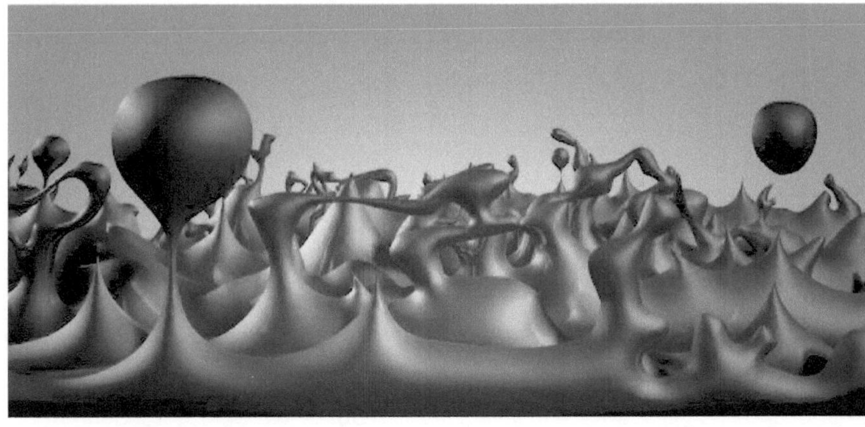

spaghetiffy the body of a traveler and also the immense heat of the accretion disk may collapse the mission. To visit a black hole that has less tidal influence we need a supermassive black hole which is infact very far away from us. Black holes just play with spacetime as we used to play with the paper when were kids, it swallow it, squeeze it, switch the individual roles of space and time in spacetime and also break it at a certain point (singularity). Thus sometimes it becomes complicated to imagine what the events would be perceived like at such strong curvatures and break down of spacetime (the singularity).

How does gravity escape the black hole if nothing really can?

A Black hole (The beast) under whose influence even the light cant escape. Things which had their way inside it could never find their way outside. Surprisingly, even the information which any object (which got inside the blackhole) ever carried gets lost inside the black hole.

Then how can a black hole's own gravity escape it?

According to general relativity, gravity is basically the acceleration that originated from the spacetime curvature. Closer the one goes towards the centre of the black hole stronger fot him the gravity and thus spacetime curvature becomes. Any object that gets into the influence of the spacetime curvature of the black hole, does not operate directly to its centre of the black hole to get that particular acceleration which it has. The gravity it experiences is just the effect of the local spacetime curvature and not the global one (at the centre). Although this local spacetime curvature somewhat is indirectly linked to the global curvature, the gravitational effects one will face by going in the vicinity of the black hole is purely the result of the spacetime curvature near that region (local spacetime curvature). Thus gravity (of the black hole) really is not what it comes from what is inside the black hole, but rather due to the local measure of spacetime curvature which tends toward the singularity of the black hole.

How does that work?

Let's suppose the black hole which we considered is the residue of some dead star. At the phase when the dead star is collapsing to the singularity and thus as its density drastically changes, it creates a local finite spacetime curvature before its collapses to the singularity. This same local curvature as a star's surface resides to the singularity becomes extremely wild until it becomes infinity at the point we call the spacetime singularity.

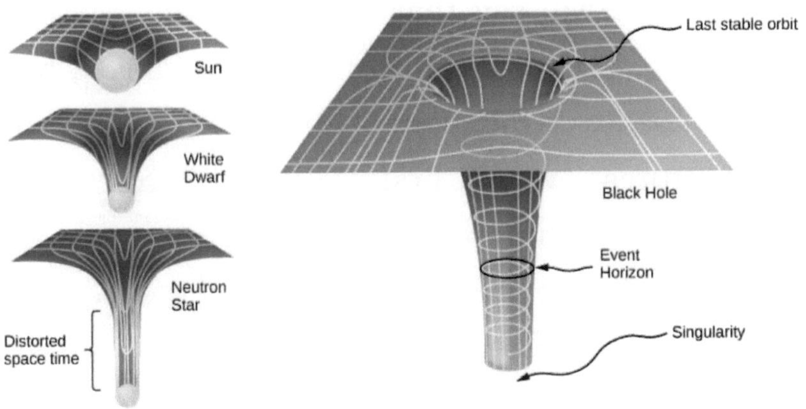

Local spacetime curvature tends toward the singularity and not the opposite!

I can't really comment on the spacetime curvature at the singularity of the black hole as we don't know what it's all about it there. Roughly speaking, the black hole spacetime curvature starts with its singularity which in fact has the most intense spacetime curvature, an effect of which it has on the nearby fabric of spacetime and again an effect of which to its nearby one and so on until finally, it reaches us (outside the horizon). This changing spacetime curvature is continuous in terms of change per distance in the vicinity.

One will say that gravity is still escaping from the horizon point to the world outside, but still no! As I mentioned above that the local curvatures were first created and then singularity. Thus one should look in a manner that while the formation of a black hole, the local curvature tends to the singularity and not the opposite! Thus the information of individual local curvatures (say) is going inside the black hole and not from inside of it!

How does black hole govern the future?

A black hole is not only a hole in space but rather a hole in spacetime. Thus not only space but time itself also is what flows inside a black hole. We have all of our world lines intersecting at a common point (The big bang). A deviation in our world line is what we define as acceleration. The mass can deviate world lines and thus make us feel an acceleration which we call gravity. If the big crunch theory is right then every element in this universe have its world line intersecting at the big crunch singularity even now.

The Penrose diagrams

In simple words, the Penrose diagram is the shrinking of the very very large scale spacetime such that the distance-time graph of light is a line always inclined at 45 degrees to the vertical, thus anything that accelerates in the has a distance-time graph always less tilted than 45 degrees as the speed of light is constant in our universe. Although we can remain rest in space , in time we cant! In the Penrose diagram time advances in the vertical direction. And our path in the diagram is called a 'WORLD LINE'. This is what a simple Penrose diagram looks like :

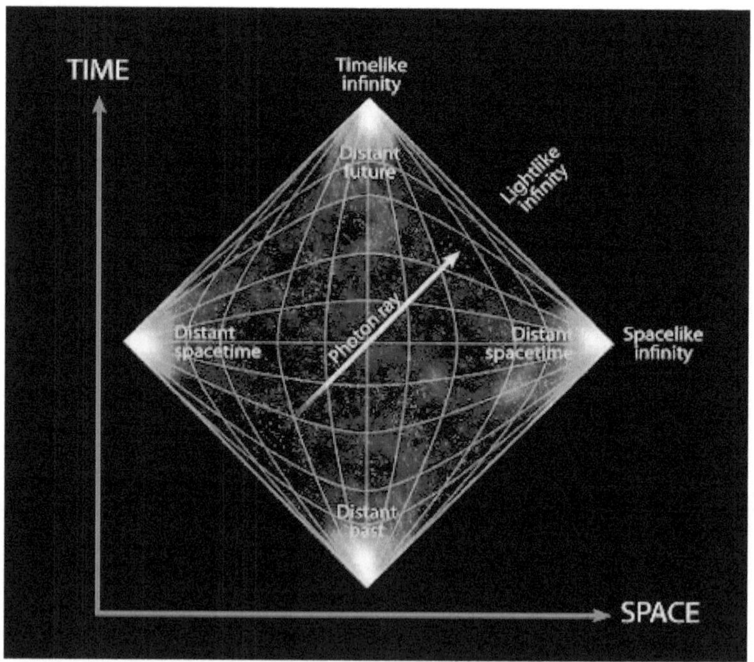

The Penrose diagram

How does a Black hole govern the future?

Imagine a black hole that is swallowing a set of world lines, including our friend astronaut's. Let's say the change in her world line occurs in a way that he barely notices it in any form of acceleration. This brings a gradual tilt in the worldline of an astronaut. When this tilt occurs by 45 degrees, then it is impossible to escape from the black hole, as one needs to boost off at the speed of light to escape it. Once the black hole tilts the world line of an astronaut by 45 degrees, it has a complete handle on his future. This occurs at the boundary in spacetime which we call an event horizon.

Is our future governed by the black hole now?

Yes! Indeed it is . In eternity that is what is only supposed to happen. All the galaxies are gonna collide with each other to form a supermassive black hole, these SMBH after that will collide to form a hypermassive one. And all the black holes including the hypermassive ones in this universe will combine to produce an ultramassive Black hole. We can't really feel this governance of a black hole as it is governing and tilting our worldlines by a very very tiny measure per unit of time. This tilt is probably unmeasurable by any of the instruments available to us up to date. But remember, the last ultra massive black hole of our universe is what governing our future until it turns into an unavoidable situation.

How can we escape from this governance?

As one nears the black holes the tilt in her world line becomes more apparent and less gradual, which can make one feel an acceleration. This is probably only the point when we get to know that a black hole wants a complete handle of our future to swallow us. To avoid this (when the observer is away from the event horizon) we have an option to blast our rocket and thus tilting our world line in an opposite direction. At the event horizon, we have to boost our rockets to the speed of light which make our mission impossible to accomplish.

Are we inside a Black Hole?

Our universe had its origin from the point called as singularity. The same thing is what we find inside the black hole . But, we don't know how shall we describe it. Quantum mechanics says that quantum foam is what dominates at the singularity. Mathematically singularity is infinitely small in terms of dimension, but we don't know what infinity is (although quantum mechanics says that Planck length is the smallest achievable length).

The only difference between the cosmic and black hole singularity is that the black hole singularity is attracting, while the big bang singularity is the expelling one. Still, there is no problem as we have a time reverse form of the black hole which we call a White Hole. Although we have never seen a white hole, its existence is theoretically possible in our universe.

Our universe is four-dimensional in nature and its shape is well described by a system of numbers which one assigns say at any point to get a curvature at that particular point, we call this system of numbers collectively as the spacetime metric. This metric gives us an idea of the possible shape of our universe. If one analyzes the evolution of spacetime inside a black hole or its time-reversed version (a white hole) then it seems very wild as compared to the spacetime evolution in the process of BigBang (when one compare). Currently, our universe is very accurately described (locally) by the FLRW metric (Friedmann–Lemaître–Robertson–Walker metric). Let's say our universe is inside a Schwarzschild black hole (a black hole with zero charge and angular momentum). Although the FLRW metric is different from the Schwarzschild metric, one can find a metric similar or even the same as the FLRW metric inside the black hole's Schwarzschild metric, but only locally ! . Then It will be a really tough task for an observer inside the black hole's (time-reversed) FLRW subdomain to distinguish the metric from the regular universe if it was not inside the black hole. Even now we cant claim that we are not in the FLRW metric subdomain of the larger Schwarzschild metric inside a black hole . Well, the spacetime evolution in our universe seems to be very smooth according to the cosmic microwave background (CMB) and in the case of the black hole, it's comparatively violent. Although we approximated a local FLRW model in the Schwarzschild model of a time-reversed black hole in which we might exist, that approximation was local ! . At the global scale say from the data from the CMB the observation does not commute with the black hole metric. Thus probably we should not believe in that we are inside the black hole. So probably we might not be residing inside the black hole (time-reversed)

What if a black hole hits our planet?

A black hole smaller than our planet is approaching us at a very high speed say 27 km/s. If it's big enough then it will swallow the entire planet. But say a black hole of a size of a hydrogen atom is set to hit us, then how would it be like to evidence?

What Kind of Astronomical Object A Black Hole Is?

Unlike a meteorite which is solid and say one can touch it, black holes are the different ones. They are basically the infinite mass-energy density in spacetime and if one tries to touch them then it is very obvious that one will either get inside it or be spaghettified, so no one can really touch it and even if you touch it by going inside it then you can't really tell that to the world outside it. And the only materialistic thing you may encounter before getting inside the black hole is its accretion disk. Every black hole is accompanied by the accretion disk around it. An accretion disk is basically a highly dense plasma revolving around the black hole. As this plasma gets near to the black hole its temperature drastically increases, as a result of which the accretion disk emits high energy radiations throughout and outside the disk. For a stable accretion disk, the radiation pressure due to this high energy radiation should counterbalance the force of the gravity, this means that the black hole has some limit on how much food it keeps along with it in the form of the accretion disk, this limit is what we call the Eddington Limit. A black hole provided with food more than its Eddington limit will just vomited out by it. For a black hole of the size of a hydrogen atom with a stable accretion disk, it requires the radiation pressure as much as it will glow like the 100 Hiroshimas to counterbalance the gravitational pull.

What If A Black Hole Of The Size Of a Hydrogen Atom Hits Our Planet?

A hydrogen atom-sized black hole as bright as 100 Hiroshimas is approaching our planet as one can now imagine how luminous that atmospheric entry would be. The black hole is set to hit earth at a whopping speed of 27 km/s. During first contact with the surface of our planet, some of the planet's mass adds up to the accretion disk and thus feeding the black hole and making it vomit outside in form of radiation pressure which digs the land inside until it exits out from the opposite end. And thus creating a tunnel throughout a planet instead of a crater. A crater created by a meteorite is broad and less deep as compared to the tunnel created by an impact with a hydrogen atom-sized black hole.

Have We Encountered Any Events Like These In the Past?

Well! the black hole moving across the earth at a speed of 2 km/s will get an earthquake throughout the globe or most of the part of it (due to violence caused by it in the interior of the earth), up to date we haven't encountered the situations like that. But It's hard to tell whether that happened in past or not. So next time, if you witness a very bright asteroid making its way from way one point on the planet to the opposite , then that is not an asteroid but a collision of our planet with a tiny black hole .

How To Create A Horizon Around Yourself?

What is the horizon exactly?

The particle horizon and the event horizon are the horizons we have studied until. The word horizon in spacetime means a region in or after which there is no way any information inside can communicate with us. The idea of creating around yourself is to create a region in spacetime where no information other than your immediate vicinity in spacetime can communicate with you . For any event to communicate with you, it must have to be in the range of your past lightcone. And you always have an option to change the orientation of your lightcones by boosting off the rockets. The shift in past light will cause the events in the past which were supposed to communicate with you before the shifting of lightcone uncommunicable.

Thus the events you observe in the not immediate neighborhood, and the stars you see in the night sky are evident to you just because their information lies in your past light cone. Once you deviate your past lightcone, that information would then fail to occupy your past lightcone and thus remain unevident to you. This is equivalent to creating a horizon around yourself that hinders the communication of the information which was once under your past lightcone but soon will not be as you tilt your lightcone by boosting your rocket. This horizon is called a Rindler Horizon.

Rindler horizon is a relative concept. If I and you boost our rockets simultaneously and maintain them at the same speed and acceleration and are in the immediate neighborhood, then the lightcones deviates by the same measure for both of us. Thus events would be common for both of us. Even now we are surrounded by a Rindler horizon. As we are under the action of gravity which has deviated our light cones since the day we all were born on this planet . And thus kind of creating around ourselves the Rindler horizon.

The Rindler horizon is something we all have ourselves surrounded with. Even if you find a void in space where spacetime curvature and thus deviation of the lightcone is minimal, you still would have surrounded yourself with the Rindler horizon as we don't know what is the shape of our universe which technically has an effect on the tilt of your lightcone at a very huge scale.

The Black hole Information Paradox.

The quantities like mass, charge, and angular momentum are the universal conserved quantities. When they get inside the black hole, they get a cut off from the rest of the universe, but not their mass, charge, and angular momentum as per the conservation law. In simple terms, you get some rotating body having a particular mass and charge associated with it and drop it into the black hole and what happens is these fundamental quantities pile up over the black hole, increasing its rate of spin, mass, and charge.

The black hole its mass from the things it swallows and also loses the same mass via the Hawking Radiation. The hawking radiation is mostly the photons, which have no information about the object dropped into the black hole. This makes us (In General) question our belief about the conservation of information. This is the Black Hole Information Paradox. Given the initial state of a system we judge its past by the imprints it has on itself as well as on the surroundings and provided there is some information loss, we are judging the system wrong! In fact, the existence of that object is itself an information. The black holes do not even leave that.

Ok! take an object remove from it, its mass-energy, charge, angular momentum, and now what is left? No existence! Everything is made of mass energy 'at least ' in our universe, it's something which even lies at the tiniest scales of spacetime. In fact, it's the rule of our universe that anyone or anything residing inside it should possess some mass or energy. Mass and energy are the basic fundamental quantity that everything in this universe has. If I tell you to remove mass energy out of any object. A stupid will probably try to burn it but understand you are just breaking the chemical bonds inside it. An intelligent student will say blasting a nuclear mass in a nuclear bomb and you have a part of mass being converted into energy, that feels nuts (Man read the question again!). But a genius will tell to throw that particular object out of the universe. Yes! That's what our vacuum cleaner (black hole) does, it sucks not only the matter, and light but also the mass, charge, and angular momentum and keeps it along with it as per the conservation of mass, charge, and angular momentum.

So next time if you want to go out of our universe or existence (equivalently), then one has to visit a black hole, which will get from you your mass, net charge, and angular momentum, and thus now you are ready to go out of the existence.

Once an object is gone out of its existence, then so does its information! Once you are out of the universe then the normal physics and its constraints, and laws just fail over there to describe the unimaginable, unexplored reality over there.

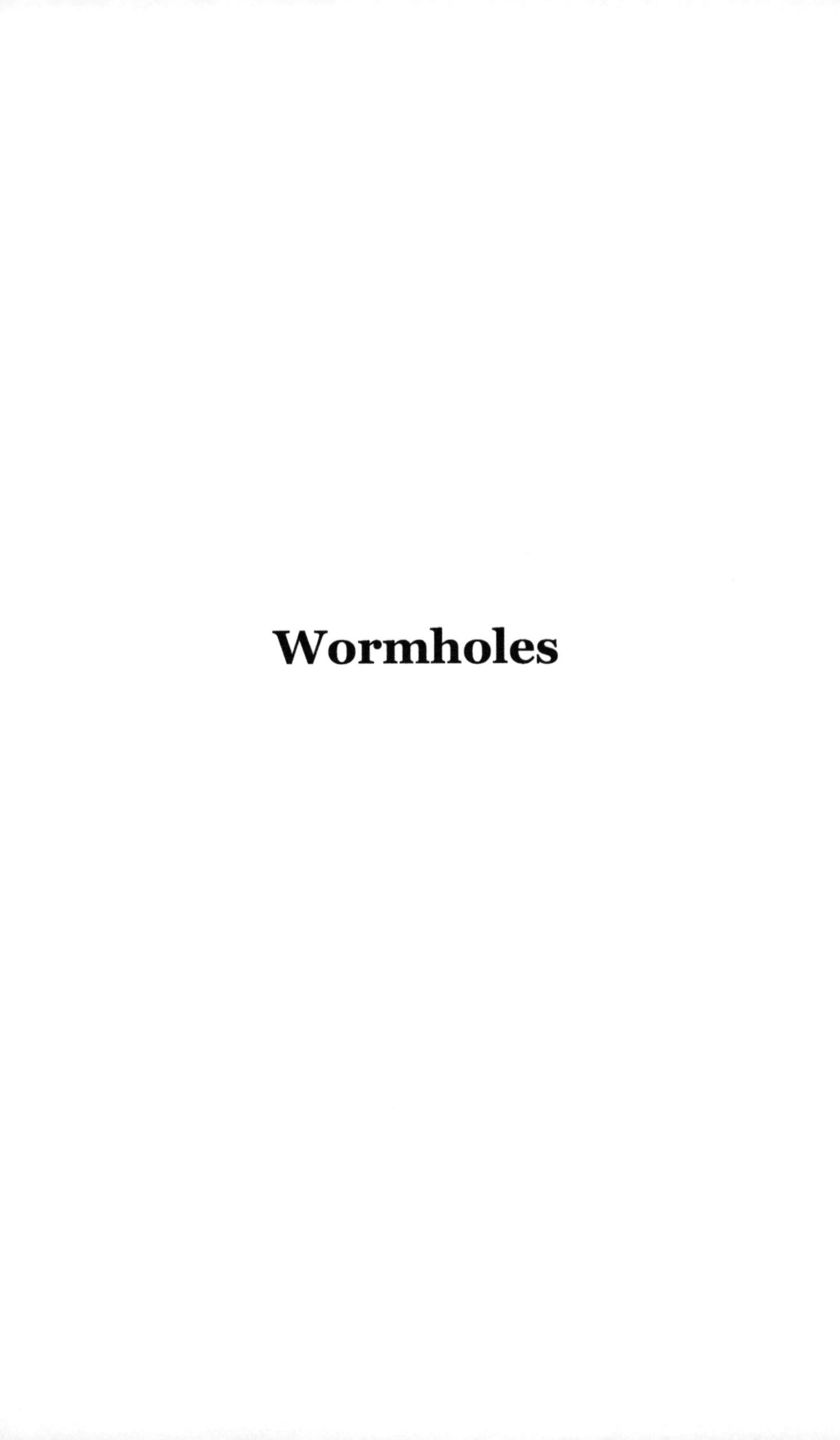

Wormholes

Einstein Rosen Bridge.

Einstein Rosen Bridge is a bridge which connects two points in spacetime. This bridge is famously known as the wormhole. Imagine a sheet of paper having points A and B on it, now instead of covering the distance straight a way between A and B what if you just fold it in a way as given here,

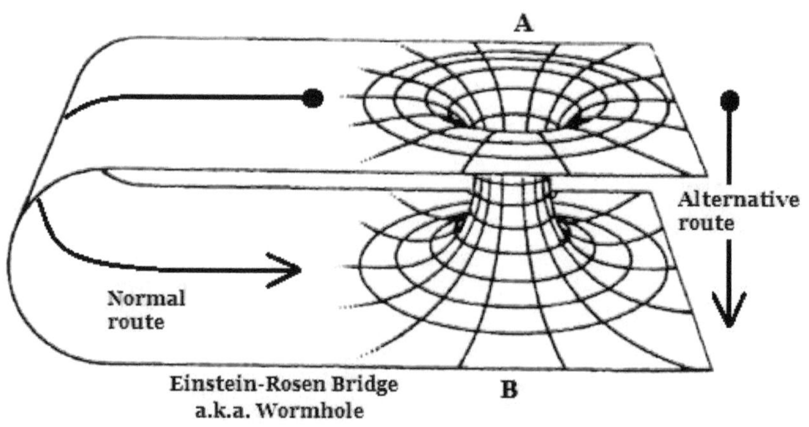

Einstein-Rosen Bridge
a.k.a. Wormhole

This structure significantly cuts down the distance and time of travel. Now this paper in reality is the four-dimensional spacetime. So it's not only the connection between two points in space but the events themselves.

For forming the wormhole one needs negative spacetime curvature which requires negative mass-energy density. Am I talking about mass which has negative value? Yes! A negative mass or energy or exotic matter is needed to maintain and create the wormhole. Can a wormhole be used as a bridge between two universes? Yes! Suppose we have two universes A and B one is running forwards in time while the other is running backwards. A black hole in time reverse universe can be thought of as a white hole and connecting the singularity of the black hole of universe A (moving forward in time) to the singularity of the white hole of backward running universe B, one can create A notch (wormhole) between two universes.

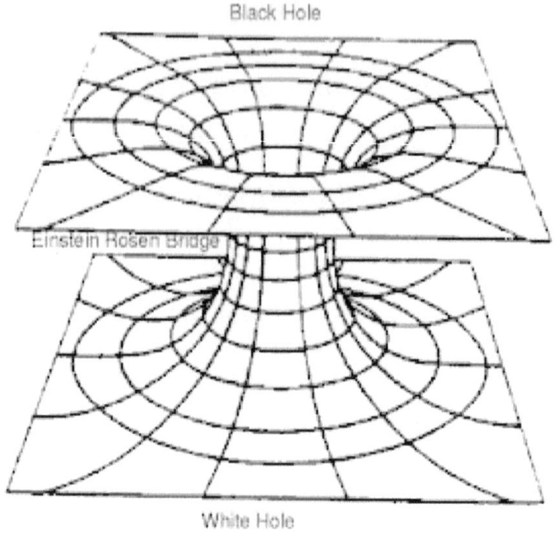

But until we haven't detected any proof of the existence of such a universe having reversed temporal flow so in that sense it remains just a theory. Why are we physically yet not able to create a wormhole? Just because we have never detected the negative mass, theoretically it can be created in the Casimir effect, but in that experiment accelerating two plates at relativistic speeds is an engineering problem and is a great challenge. A wormhole is basically a spherical hole as it's a hole in four-dimensional spacetime. How to use a wormhole as a time machine? well! take one mouth of a wormhole in any noninertial frame moving at a speed comparable to the speed of light and keep the other end at rest. Now bring back the travelling end of the wormhole near to the rest one and you have mouths of a wormhole connecting at different times. As travelling end travel at a speed comparable to the speed of light, it has its time dilated and in fact, has travelled in future and other remains in the past, thus allowing one to in the future to time travel in past. But here the restriction is that the future time traveller will never be able to go earlier than the time when one of the ends started accelerating for the experiment. Maybe that's why we don't have people coming from the future as they can travel no earlier than the time when the time machine was built. Thus here we have two ends connecting different locations in time, making the wormhole the possible time machine.

Creating the negative mass-energy density.

The initial part is introductory. One can directly jump to a wormhole or time machine part (at the end of the blog) to save time. But reading some introduction would be the best for a greater experience.

What is the vacuum or zero point energy ?

What does it mean when we say that there is nothing in space? Humans really imagine 'nothing' as is still something we are searching for what it is . 0 is the number which may represent nothing but still absolute constant may never exist and thus making a zero a nonzero . In the similar fashion there no point in universe called as 0 energy point . Our universe has thin layer of energy covering it entirely. This energy is called the vacuum energy, which is present all over the universe. According to the third law of thermodynamics, absolute 0 K is untenable and that is the universal constraint. Provided we have vacuum energy over the entire universe, this law becomes very fundamental.

What is the origin of the vacuum or zero point energy ?

Virtual particles pair annihilation is the source! This process is what occurs throughout the universe.

To elucidate, at very small scales of spacetime two pairs of particles are created one is a normal particle and the other is its anti-part. One can say an electron and positron created themselves out of nothing. Once these particles are created they leave behind some negative energy density which in fact returns to a positive value as soon as virtual particles attract and annihilate each other. Strictly speaking, this doesn't violate the energy conservation law. The time interval between the successful pair production and annihilation remains very small. Heisenberg's uncertainty principle says that the change in energy density becomes harder to measure if the time interval of change is very short. So technically in empty spacetime for very short intervals of time, there always remain a chance of change in energy. This logic behind the uncertainty principle is responsible for vacuum energy or particle pair creation.

The Cassimir Effect

This is an experiment that can be performed in a vacuum. In the experiment, we have two nonconducting plates separated with a very small distance between them (say at the scale of a nanometer) and of course we have the very basic vacuum energy or zero point energy with us in the setup. Energy lower than zero point energy density is considered negative energy (obviously). The following diagram shows the setup for the Casimir effect

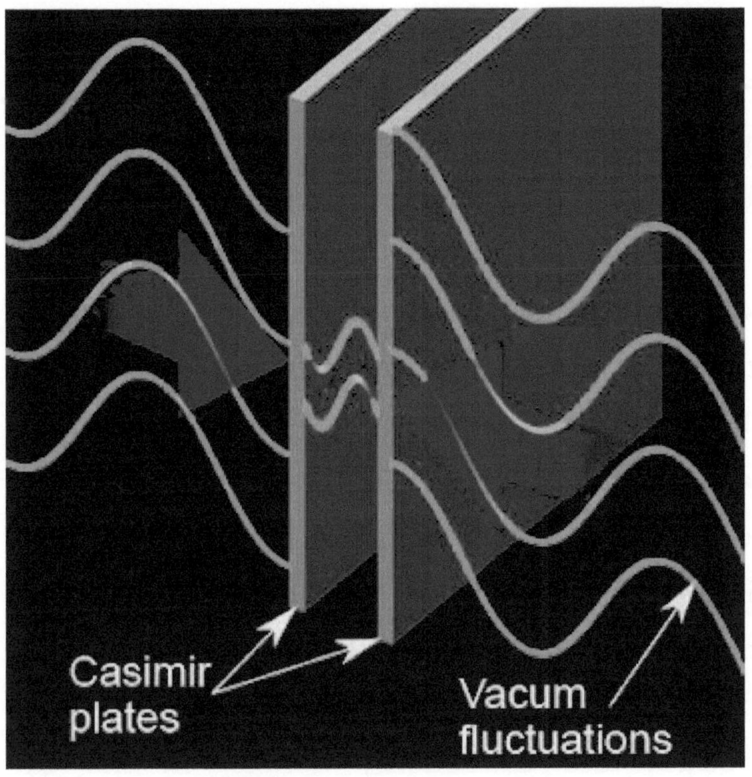

WORKING: Both of the above plates have one of their sides facing to open space and the other facing each other. On the face oriented to the open space, there is always some force acting on it, and the same for the other plate. This force is due to the virtual particle pair production effect on both sides of two plates. But in the region of the gap between these two plates the condition is a bit different than the faces oriented to the open space. As we know matter is a wave and vice versa, the dimension of the wave associated with the virtual particle created inside the gap between two plates should always have to be some fraction of the gap between two plates. These get in some energy range for virtual particles to be created in that gap, this energy boundary is always less than the energy density virtual particles could have as a wave. Thus there is the net force on the plate which makes them stick to each other. But wait we are in the vacuum and the energy density in the gap between two plates is less than what is outside it (zero point energy !). Thus technically we can create negative energy densities in such systems.

The Gravity

How does A Gravity Tractor move an asteroid from its orbit?

Say an asteroid larger than a football ground is heading towards us , so is that the end of the world?

Well, that depends on the factors of what the asteroid is made of, its angle of collision, the mass of the asteroid, and the velocity of the asteroid. If the asteroid is made of various metals then it is more hazardous than the ones which are mostly composed of carbon. And if the collision is taking place say at an angle which is less than 90 degrees then the kinetic energy (energy possessed by every moving object) would be wasted in peeling off the surface of the earth and less violent shockwaves would be created and when the angle of collision is 90 degrees then the most of the kinetic energy of the asteroid is transferred in the form of a shock wave and rest in the form of heat when it penetrates the earth surface. When these shock waves make their way to the oceans then tsunamis will get triggered (flushing out a whole city or country or even the entire world). I should describe the role of mass and velocity collectively as the momentum (= Mass x velocity) is what plays the role here. When the asteroid hits the earth its momentum gets transferred to the earth. I guess an impulse (momentum imparted in a very short interval of time) would be the great replacement of term momentum over here. The entire humanity is in danger. The threat of such asteroids always remains (Although less probable but may or will happen).

How do we detect the asteroids?

Look up in the clear night sky until you watch any falling star and that is how we do it! These falling stars are just meteorites which are just the chunks of the larger asteroid or any comet. Asteroids really don't have their own light so unless the rays of the sun fall over the asteroid and some traces out of it get reflected, we are not able to target them. And to check for its size, shape, and orbit we use radar technology. Once the trajectory is calculated, one can predict the future path of an asteroid.

How can we shield ourselves from these hell stones (real ones! literally!)?

Asteroids may move at velocity as high as 30 km/s and even more! They have violent trajectories and irregular surfaces. These factors make asteroids harder to detect. And the best moment to tackle any asteroid is only when it's more than 7.5 million kilometers away from us otherwise they are considered potentially hazardous ones (provided they have enough mass and velocity to create massive destruction) which are even harder to deal with if we aim to deflect it. But how to deflect an asteroid off the track? This how :

By Nuking it:

Yes in a situation when an asteroid (potentially hazardous one) is calculated to hit the earth, we can nuke it to either move it to the safer track or to destroy it. It depends on the mass of the asteroid that how many nukes should be detonated to move it off the track (according to Newton's third law). But still, the problem is that even if one turns an asteroid into pieces by nuking it, the meteorite or micrometeorite formed due to the destruction of the asteroid may destroy our satellites. Thus there is a huge loss in the economy. But still, human life is important. It's an effective way to destroy or boost the asteroid off the track.

Deflecting an asteroid by the light :

If I tell you that you can deflect a metal sheet or any object by just making some light incident over it, then probably you may say I am kidding! But no! I am not. Light imparts the momentum like a normal traveling object. But light has no mass! But it has energy and as energy is the same as mass, light has its own momentum to impart. The pressure generated by the force imparted by the light due to its transfer of momentum is called the Radiational Pressure. An asteroid can be moved to some other track in this way. It's a good way of deflection as all we need is just high-energy light rays to impart great momentum over an asteroid to deflect its path.

Deflecting an asteroid by a gravity tractor :

Two masses always have the property to attract each other by the means of gravity. Let's imagine an asteroid (potentially hazardous one) moving towards the earth and we try to deflect it by means of some mass by having a gravitational tug between that particular mass and an asteroid. This mass is basically a moving spacecraft with an ionic propellor sent by us which is set to hover near the asteroid. The spacecraft is placed in such a way that its gravitational pull on the asteroid is perpendicular to the plane of the asteroid belt. If the spacecraft and an asteroid if considered as a system then a slight change in spacecraft velocity will affect the spacecraft and asteroid system's momentum as a whole, moving it off the asteroid orbit plane. The force due to gravity due to spacecraft may be very less on the asteroid but over the years the consistent pull can drag it from its orbital plane and thus securing the existence of our planet earth or may be of the humanity as a whole.

The Lens Made Out of Gravity.

Every mass has a spacetime curvature associated with it and when light bends by some angle say when it gets into this spacetime curvature we say that the light is being lensed by the lens of gravity, I mean a gravitational lens. This technique is what Arthur Eddington had used to verify general relativity. This experiment was conducted during a solar eclipse for background starlight to get visible . In a nutshell, the experiment was to calculate the apparent shift in the position of stars when they are accompanied by the sun's gravitational lens. The result was that the stars had their position shifted from the normal. Newtonian gravity could never have predicted it.In gravitational lensing there is not only the spacetime curvature that affects the path of light but also the frame-dragging effect. Frame dragging effect is basically a drag in spacetime that results from the rotation of the planet or any star or a black hole. Due to the combined effect of the spacetime curvature and the frame dragging effect one may get to see the interesting patterns of galaxies and stars whose light encounters this lensing. An interstellar object, say which is beyond some star or planet, due to which it's normally not visible to the observer, may also start getting visible when it gets into gravitational lensing as given below:

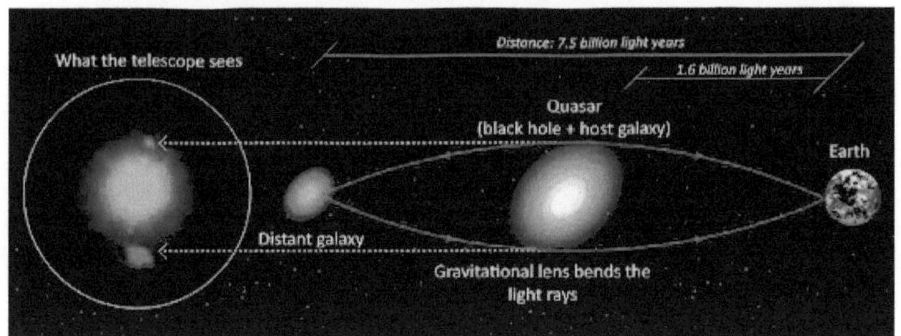

Image by the james webb telescope

Here, the distorted image of the galaxy is due to the lensing by some interstellar object.

Holes are black, making them hard to identify from the surrounding dark universe, but the effect of lensing they have on a surrounding light of a star or a galaxy makes them detectable. Gravitational lenses can be used as the agents of the telescopes to see distant objects. Also, it reveals the other side of an interstellar object which one cant normally see without lensing.

Waves through the fabric of Reality.

The universe or spacetime is a 4-dimensional entity. Gravity is the curvature in spacetime, which we feel in the form of acceleration. The waves in the fabric of space-time are what we called as gravitational waves. Imagine spacetime as the stretched rubber sheet and the waves through it are the gravitational waves. Remember that gravitational wave is the curvature in spacetime, so curvature in time is also the main factor to consider. Spacetime is connected to each other, one can't really have his time dilated keeping his space coordinates constant and vice versa. A gravitational wave if passed through any object has some stretching and contracting effect over it. The effects of gravitational waves are apparent only at large scales. Multiple waves producing events occur in this universe at any given time, information granulated by the patterns of stretching and contracting produced by these waves can even be encoded to the information of the source itself. We have a LIGO observatory which is located in Louisiana in the united states, LIGO has detected 90 gravitational waves since the time it was established. If a Gravitational-wave passes through this L-shaped system one of its two arms oscillates in length which shows up by altering the path lengths of two light beams which makes two beams fail to destructively interfere with each other and flashes of light can be achieved on the detector. On September 14,2015 LIGO detected its first gravitational wave, which had two colliding black holes as their source.

Laser Interferometer Gravitational-Wave Observatory (LIGO)

There are many sources of gravitational waves in our universe for eg. the colliding neutron stars, merging black holes, colliding stars, and many more. This is how two neutron stars produce gravitational waves :

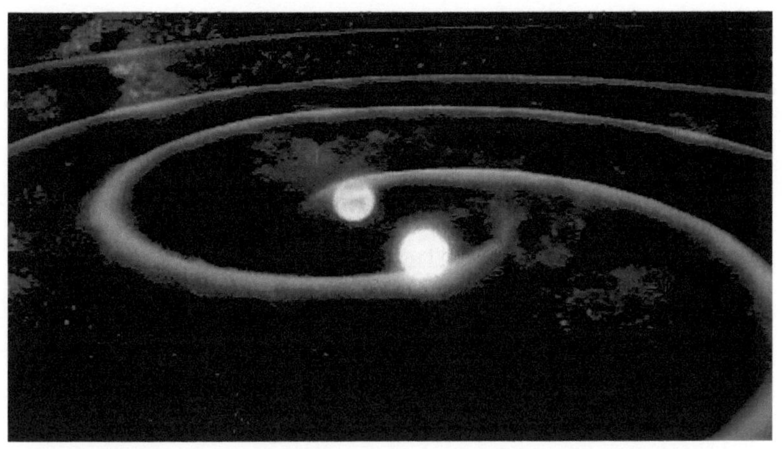

Two tend to collide neutron stars producing the ripples in the spacetime (gravitational wave)

These waves transcend throughout spacetime at the speed of light or smaller and surprisingly even greater. Suppose we have a bulk being (A multidimensional being) that sends a wave through spacetime at a speed that is more than the speed of light. Then this can happen! Wait! speed of light is the ultimate speed limit. To restate, one cant really travel faster than light in 'our universe'. Thus the universe is violating its own law is all ok! These laws are in fact made only for the mass-energy that reside in it which is absolutely ok and we can say laws of special relativity remain unviolated. Well! the gravitational waves we have detected were all traveling at c . Gravitational waves are actually the 4-dimensional waves, the above image is just the two-dimensional analog of it. If our universe is one of many universes in the multiverse, suppose some other universe collides with ours which may result in the formation of gravitational waves. Such waves would be very interesting to encode and also may prove the existence of a multiverse if there is any! Measuring gravitational waves is a complicated task to perform as there always remains some dampings that alter the measurements.

Quantum Mechanics

Quantum Mechanics (Introduction)

Let's start with the word quantum, it suggests one about quantity or something which is discrete . Quantum Mechanics describes all four fundamental forces of this Universe with this discreteness . Quantum Mechanics basically deals with the quantum theory which explains the behaviour of particles at the subatomic level. Well! what if I tell you that an electron can pass the wall without crashing on it? Is that magic? No, it's the quantum physics. At the sub-atomic world, it's very common for such things to happen. In classical physics, everything is well defined, for you to cross a wall without colliding you may have to wait for billions of years. In classical physics the total mechanical energy is given by kinetic energy + potential energy, in the case of subatomic particles we have the Schrodinger equation, it's basically the quantum analog of the classical total energy equation. It gives the values of the most probable potential and kinetic energy. At the quantum level the parameters like position unlike classical physics is a very weird concept for example in quantum physics an electron can simultaneously be at the two positions . Such particles are called to be in the superposition state .

This can be thought as the most intelligent picture in the world, each and every one of the above scientists have significantly contributed to the development of quantum mechanics.

Scientists have proven through various experiments that light is a wave. And by the photoelectric effect, light shows the properties of a particle, how can a wave be a particle? unbelievable! And if a wave is a particle can a normal world particle be a wave? of course! It has experimentally been proven that a moving particle behaves like a wave in some conditions and like a particle in others. In Quantum mechanics we have the de Broglie wavelength for it, which connects the momentum of the particle with its wavelength itself. A subatomic particle can never be at a definite position or at an absolute inertial frame, as small quantum fluctuations are always present in this universe, this energy never allows subatomic particles to be at absolute rest, thus they always have some wave associated with them. The wavelength of each and every particle has its reach throughout the universe. In quantum physics there are certain rules and conditions for a particle to follow :

1) Heisenberg's Uncertainty Principle: The more precisely you measure the momentum of the particle, the more uncertain you are with its

position and vice versa. Mathematically can be stated as (One can skip the mathematics part) :

$$\Delta\chi\Delta\rho \geq \frac{\hbar}{2}$$

here, delta x is the uncertainty in position

delta p is the uncertainty in momentum

2) Pauli Exclusion Principle: No two electrons can attend the same quantum state. This principle if tried violated may give rise to the external pressure . For example, the electron degeneracy pressure which maintains the white dwarf from collapsing into the neutron star. This pressure is not due to the underlining new fundamental force, but it's the mathematical tool or an approach of the universe to not let the star violate the Pauli exclusion principle.

Sending information faster than the speed of light is also possible by the famous quantum physics phenomenon, also known as the quantum entanglement or The spooky action at the distance. scientists have also physically teleported the photon from one place to other by the same principle, imagine same with the normal matter or with human beings. That's incredible!

Quantum physics also explains various astrophysics phenomena which classical physics and even general relativity cant explain. It also explains the mechanism behind the four fundamental forces of the universe as being accompanied by force-carrying particles like photons, gravitons, gluons, etc . For Quantum mechanics to be a complete theory we need some experimental evidence.

Quantum Mechanics has its approach in fields like physics, chemistry, biology, engineering, artificial intelligence, computer science, and many more. Quantum technology will be a dominant sector in the study, technology, development, and employment in near future.

The Schrodinger's Cat Experiment.

Quantum mechanics or the mechanics going at the subatomic level is where all the parameters like position, energy, momentum, etc are not distinct (contrary to classical mechanics). Every particle has a wave associated with it. The smaller the mass the particle has, the larger its de Broglie wavelength would be (provided that the velocity of the particle is kept constant). Suppose we have a wall as a barrier for an electron to go to the other side, seeing the electron as a wave, in that case may allow the electron to get to the other side like a ghost! That is quantum tunneling. If a particle has its wavelength spread beyond its barrier then there are significant chances for a particle to tunnel the other side. In the nuclear decay process of radioactive atoms, the alpha particle needs to get past some coulomb barrier to get radiated. In terms of classical mechanics, the particle cannot escape the nuclei unless it's being provided some extra boost of energy. Quantum mechanics says that the alpha particle has a non-zero probability of escaping this barrier (by tunneling) and that is very true otherwise alpha radiation would not be as common in our universe as it actually is.

Schrodinger's cat experiment (thought experiment):

Lets have a physicist we perform this experiment by having a cat , the nuclear reactor and the poison bottle as the apparatus in ant box . If the nuclear reactor reacts successfully by radiating an alpha particle (which is a matter of probability) then the nuclear energy is converted to the mechanical energy of a system which shatters a poison bottle which kills the cat inside the box . Lets say here we have a fifty-fifty percent chance of the nuclear reactor reacting successfully. If a nuclear reactor reacts then it converts its energy into mechanical energy which is associated with some setup to shatter the poison flask present in the box from which the cat inside the box will die and if it fails to react then the cat remains alive. So here in this experiment, the very quantum mechanical phenomenon of tunneling is translated from the subatomic (microscopic) to a macroscopic scale as the fate of a cat. Let's start the experiment by having the box closed. If I ask the physicist about the status of the cat inside the box then he will tell me a cat is dead as well as alive ! . To shortly divert from the topic , When we express say a position of a quantum particle by a wave, then that outcome of probability is not in terms of 'or', it is all at a single instant result, thus it can be stated to be acquiring multiple positions at a single instant in time, it can also be said that a particle is in the superposition state. Returning back to the experiment I would say that the alpha particle has been radiated as well as it's not (the event is in the superposition of radiated and not radiated), thus it can be said that a cat is dead as well as alive. But of course, as soon as the phycisist open the box there would be just a single outcome . What leads to the collapse of this wavefunction? It's the observation. Observation collapses the wavefunction. Can we say a conscious observer instead? Yes! Is there a moon if no one is observing it? are some of the famous questions which come to mind if one gets into this theory more deeply. Let's say a guy also famously known as a wigner's friend is observing (through a window) the physicist who was conducting the above experiment, so the status (imagine room as a box) of the physicist is what is determined by the wigners friend as an observer and thus there lets consider a one more observer who is observing wigners friend (let's say) who is still in another room, deciding his status and so on. This demands a universal consciousness or a universal observer who collapse the wavefunction of status of each every inhabitant of this universe . Many relate this universal observers to be The God. Well! we don't know whether it's a god or anyone or any consciousness which is universal out there. All the thing to say is that this theory raises such questions which somehow may convince us regarding the existence of the universal consciousness.

God may also exist in extra dimensions say in 10 or 11 dimensions. But I guess still the existence of god remains the question unless and until humanity has solid proof of that.

The Observer Effect

The observer effect

A state of a system remains undefined unless it's observed. In quantum mechanics observation collapses the wavefunction, where the wavefunction represents the incompleteness in our knowledge of the data of the observable. In Schrodinger's cat experiment we saw how this observable at the micro level affects the states at the macro level. In that sense, our universe is merely the outcome of our observation. This is the Observer effect.

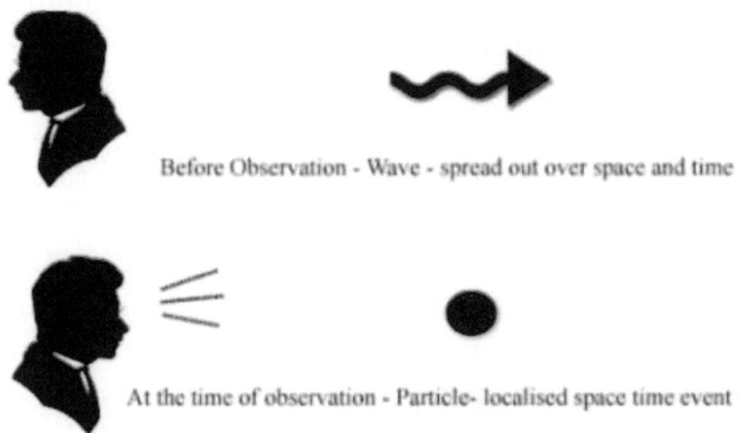

What does the observation mean in the observer effect?

The observation is not just constrained to what we see but rather in general to what we sense. The basic observation which decides the fate and decided the past of our universe is the sense that conscious observers exist. The formation of life itself to discuss is a very very finely tuned event. In the same event, we also had mass extinctions and the ice ages. The four fundamental forces of our universe and their strength are so finely tuned to allow our existence. One can imagine that the value or the strength of these fundamental forces before the time they originated was just in the form of one of the outcomes of the wavefunction and the observation of our mere existence collapsed this wavefunction and insisted the four fundamental forces of our universe to achieve specific values so that we the conscious observers exist.

Is there a moon when nobody looks?

Yes, there is! Moon has a great role in life formation. Thus even if we are not looking at it, we feel it, and sense it in terms of our existence (overall and now itself)

What I am proposing might seem very baseless but remember what I always say is 'we humans tend to see the entire universe as what we face in our daily lives is ' which is not right ! . For eg. Some people on our planet believe that the acceleration of gravity due to the earth (9.8 m/s) is universal. This notion is completely wrong. In reality, what we see around us is just one of the conditions or events of the entire universe, and looking at these conditions to apply throughout the cosmos would totally be wrong. The human mind is developed in such a way that we believe in discreteness why is it so? just because our mind is designed to think that way. Imagine wherever you go you will collapse the function by the senses you have then why would we as human beings will be believing in the state which insists not on one but the combination of the events, This event is something we can't imagine. As John Wheeler said 'No phenomenon is a real phenomenon until it is an observed phenomenon.'

The Quantum Zeno Paradox

Quantum Mechanics is the language of probability. In quantum mechanics, events in the future are described by the probability wavefunction. And the final outcome is what observation or measurement decides.

Does observation refer to human observation particularly?

It can be, but not necessarily as it can be a detector or even the dependency of one observed system on something unobserved which makes its wavefunction collapse.

The Quantum Zeno Effect.

Let's say we have an event where a whole event is described by some sort of wave function. That event be suppose an electron shooting out of an electron gun to the metal plate. And let's say (In this case) we have only two results say an electron at initial position A and a final one over the metal plate let's call it B. Quantum mechanics describes this event in terms of probability. And lets say here we have a 50-50 chance of an electron being at state A and at state B.

Let's say we observe an electron at state A or simply the wavefunction collapsed to point A. Then what after that? To go at point B the wavefunction has to spread again so that choices (Point A or Point B) can again be made so that an electron can land at point B. But what if we do not allow the wave function to spread out by continuously observing it? I mean not giving a mere chance for wavefunction of an electron to evolve from one discrete result so that it can jump to the other (point B) . This will halt the electron's motion at one particular point in space. And the motion will not evolve at all. This is the quantum Zeno effect. Which says that unaltered observation or measurement of the system halts its progress. Refer to the following example to understand the concept more clearly (Still one can skip it).

Example 1: Imagine an electron at a lower energy level in an atom. As soon as we provide no energy to that electron, it will always stay at the lower energy level. Providing no energy to the system is itself an observation in this case which halts an electron at the lower energy state. One may feel unsatisfied with this example. Thus one can have a glance through yet another example of the same system. But now take an excited state as the matter of consideration.

Example 2: Imagine that we provide some energy to an electron and by which it gets to some excited state, and soon as that transition from lower energy level to higher energy level takes place we keep providing it energy (unaltered). Then an electron will never hit the ground state again at least until we shut our laser beams off. The mere act of providing an electron the energy is the process of observation or measurement which makes its way to a higher energy state. An unaltered series of measurements makes the system remain unperturbed.The example given here was a very basic example of quantum zeno effect

Zeno's paradox in a sentence: Repeated observation of a system halts its progress.

Exaggeration of Zeno's paradox :

As Matt O'Dowd (PBS Spacetime Youtube) says, If someone shoots you with an arrow do not blink your eyes. (Here we just have two states: the initial position of an arrow and the final one when it shoots any target)

Explanation: Look at the arrow at its initial position (make a measurement) and don't allow the system to progress (quantum Zeno effect) and the arrow won't hit you. And the moment one closes her eyes the wavefunction spreads out and can take an outcome of hitting her. Although this example is overexaggerated but has the power to explain the reader what is going on in the weird quantum zeno effect.

What exactly is happening during a Quantum leap?

If you shine a laser on any atom it provides an electron inside it the energy to go to the higher states. But while attaining this high energy state, the electron doesn't physically travel between the individual energy levels in the atom and instead makes a quantum leap. One can imagine it as a kind of teleportation-like thing in which an electron has between two energy levels of the atom.

But what exactly is happening during a quantum jump or leap (whatever you say) if the electron is not physically travelling between the energy levels. Before we move further, Do we really know what motion exactly is. Many of us will say that's such a dumb question, but believe me, it's not. We, humans, love to imagine the entire universe in a way we perceive what is at our scales. But really that doesn't apply.

What exactly motion is? (In terms of quantum mechanics)

Let's take an example of a person walking in the garden. We say that she is travelling some definite distance in space, let's say in meters or maybe in cm or nanometers if you like (even if the time taken is very very small). Think! as I go on! As we go still at the smallest scales, we get at the scale at which the person's motion is considered, then it takes virtually no time! to travel say a unit magnitude of that scale. So is the person still considered as physically travelling there? (Think!). In quantum mechanics, we have Planck length as the minimum possible length. Motion if considered at the Planck scale, then one is restricted to travel no distance lesser than a Planck length physically! Thus even if a person is physically travelling in the garden, at the Planck scale she is not!

Even if you call (say) yourself to be in the rest frame, it's not the case at the tiniest scales. You are moving to say leftwards at the speed of say hundred Planck lengths per second which is although unnoticeable at our scale, but you are in motion (Technically). Shortly, there is a lower bound of how the minimum physically traversable distance. l and at the tiniest scales, we seem to travel like what an electron does in atoms (a quantum leap). So even if our wavefunction does not seem like what it is in the case of subatomic particles and is collapsed one instead, but still it has even small but some wavelength as there is always associated some uncertainty in our position at the Planck scale. Thus if any motion is analyzed at the tiniest of scales say at the Planck scale then it's not a physical travelled motion, but is the outcome of the wave nature which we possess.

What exactly is happening during a quantum leap?

This nature wants symmetry, and symmetry wants every system to be at its lowest energy level possible. Atomic orbitals where electrons stay are the most stable state if the atom is considered as a whole system. If we provide some energy to one of the electrons in the atomic orbital, it achieves a higher energy state and simultaneously makes its way to the lower energy band. We provide energy to these electrons by shining a laser over them which they absorb and make the oscillating transitions. In short, we can say that in this case, we have only two outcomes of where the electron can be when provided some energy to it, one to the lower and the other to the higher energy state. And the measurement and observation are basically the actions of shining a laser on the system. This measurement makes an electron take a higher energy state which is most stable at these energy levels that the electron has and again return down to the lower one. Physicists are still in process of figuring out if they could trap an electron in the middle of the two individual energy levels by the Quantum

Zeno effect (provided our measurements are strong enough) and thus proving the theory I mentioned.

(for those who want to understand in layman or equivalent)

In short, electrons inside the atom if viewed in wave nature then our mere observation of shining the light make its wavefunction collapse at two destinations (Higher energy state and the lower one!)

A Recap

Throughout the discussion, I got the reader to

what motion looks like at the tiniest of scales by proving to her that physical movement is not possible at the Planck scale. And also how I showed how we tend to make a transition at such scales. Then I declared that absolute rest is not possible. In the last section, I answered the main question of this blog.

An approach

Here, In the discussion, I did not generalize or relate the electron's transition to the physical motion (I did reverse). Instead, I generalized the physical travel or motion to the tiniest scales at which it does not seem like a physical journey or motion, at all, changing the reader's interpretation of what motion exactly is.

Thermodynamics

The Laws OF Thermodynamics

The laws of thermodynamics are among one of the finite laws which govern our universe. To this date, nothing has been evidenced to violate these laws. There are in total four laws of thermodynamics that unfold the fundamental nature of our universe. Below are the elaborations of three individual laws of thermodynamics :

The zeroth law of thermodynamics:

'IF TWO THERMODYNAMICS BODIES ARE EACH IN THE THERMODYNAMIC EQUILIBRIUM WITH THE THIRD BODY THEN ALL THREE BODIES ARE IN EQUILIBRIUM WITH EACH OTHER. '

To elucidate, the zeroth law of thermodynamics is a very obvious way of describing the fundamental nature of heat transfer.

The first law of thermodynamics:

' THE TOTAL ENERGY IN OUR UNIVERSE REMAINS THE CONSTANT. '

Elucidation: In school or college we might have learned about this law. The most common alternative statement for this which many have studied may be that ' Energy can neither be created nor destroyed ' which is absolutely correct. What I stated is also the same thing, it's that the total energy which was been created after the big bang will always remain unaltered throughout the lifespan of the universe. Mass is energy so the conservation of mass is the consequence of the same law.

The second law of thermodynamics:

'THE ENTROPY OF OUR UNIVERSE ALWAYS INCREASES OR REMAINS CONSTANT AND NEVER DECREASES. '

Elucidation 1: Entropy is basically the thermodynamic variable that accounts for the increase in information or randomness of a system. During BigBang, the entropy was at its lowest value, and at the BigCrunch, it will be having its highest value. It is due to this law that nothing is permanent in this universe and everything ages from the stars, and planets to the black holes and everything in the universe. In some cases, entropy also may remain constant, say when one reaches absolute zero temperature and also in many conditions.

Elucidation 2: To state more clearly, the perpetual machine which converts gets some energy input into it and which can work for infinite time is just impossible, why is that so? because of the second law of thermodynamics. There is always remains some damping to resist this perpetual motion. Even if we ignore the mechanical friction in the machine and get it in a vacuum it still won't work as at the tiniest scales of spacetime there remain the virtual particles that damp the motion.

The third law of thermodynamics:

'THE ABSOLUTE ZERO TEMPERATURE IS UNATTAINABLE.'

Elucidation: When one reaches the absolute zero temperature, entropy attains the constant value. But at the tiniest scales of spacetime, there is always some temperature associated with it, due to which absolute zero becomes impossible to attain.

Thermodynamics And End of The Universe.

" We, the humanity would leave no stone unturned to ensure our existence, unless and until our universe end itself. "

Nothing in this universe is permanent! Does this statement apply to the universe itself? yes of course! There could be various reasons for the end of the universe like the heat death of the universe, vacuum decay, etc which in fact needs another blog. We are here gonna be focusing on the end of the universe and how thermodynamics plays a role in it. So let's start!

We, humans, lightened our first torch by the fire initiated by the tapping of two stones which in fact were later replaced by the stick and flints. We got to know about the resources we should use to maintain this torch for a long time and today we are at the level of harnessing energy from our fossil fuels, the energy from our

Sun and soon we would be harnessing it from the asteroids by mining it. Our sun is continuously exhausting its fuel and the residue or by-product of which is mounted over the sun's outer layers thus increasing its size. In about 7.59 billion years it will swallow our planet. Is that the end of humanity? No! To recall our history we have made tremendous progress in STEM (science, technology, engineering, and mathematics) concerning the time, we are finding new ways to harness energy from the universe around us. There is still a lot of time for us to be swallowed by the sun, in fact much enough to develop a technology for the entire humanity to transfer to some other habitable planet. Although potentially hazardous asteroids have always threatened us, we will surely become enough capable to tackle them shortly.

We humans by our creativity, logic and the power of innovation will always manage to survive until there is no energy source to keep us working. Let's zoom into the very very far future where we humans are done with exploiting each and every energy source available to us from planet earth and our star to harnessing the energy of the entire galaxy! It's not an exaggeration as technically it's possible and is thus just an engineering problem that can be resolved by harnessing energy from sources other than earth like asteroids, some other star, or the dwarf planets. After exhausting the entire galactic fuel, we humans will seek the entire group of galaxies in this universe. After exploiting each and every resource out there in the universe we humans may seek for harnessing the energy from the dark matter by converting its mass into usable energy.

Imagine the level of technology of civilization that has exhausted and utilized the energy of each of the stars in the image which is in fact just a patch of space seen from or around the orbit of the earth!

What after harnessing every source out there? what we will be seeking, for humanity to keep burning its torch to stay alive? Well, there still remains a way where one can harness some positive energy density from the universe by creating a negative energy density as the byproduct, this would not be a violation of the second law of thermodynamics as the total energy of the universe mathematically speaking will remain constant. If I remove some energy say from zero point energy and do some work by which in fact be stored in some form of energy, this energy if added to the negative energy density created will mathematically be zero energy density, not violating the very fundamental first law of thermodynamics!

After harnessing energy from every source available in the universe say from galaxies to extracting energy from the zero-point energy. Now civilization will seek the energy which itself is driving the expansion of the universe which is the dark energy! This is a tremendous amount of energy to run out with or say it will be sufficient enough until the universe end itself.

The entropy of the universe always increases or remains constant but never decreases (To know what entropy exactly is go to the laws of thermodynamics blog). The increase of entropy is the reason why everything ages including me, you and everything out there and in fact the universe itself! Collectively, each and everything in this universe have their fate inside the black hole! But how? We are all and everything in this universe which has some mass and energy linked with each other by the gravity. We live on planet earth which will soon be gulped by the sun along with our fellow solar system neighbours (planets). Given a sufficient amount of time, every star would have sucked up its system into it. Now these individual stars has gained a lot of mass to collapse into some black hole or neutron star which will again be having their fate into the black hole at the galactic centre which again interact with some other galactic centre in the universe and so on, until each and every particle of mass-energy are inside the black hole. This ultra ultra massive black hole will be the only thing left in the universe to not consider some traces of photons. Black holes are basically the highest entropy contributors of the universe as they have a lot of matter inside them which had or still may have there own entropy. During the initial fireball or Bigbang, the value of the entropy was at its lowest point, this entropy is supposed to take its maximum value at the big crunch (end of the universe).

Once some amount of information which is in fact an increase in entropy is created it has to be in the universe forever! The universe has to carry this entropy burden until it ends into the big crunch or any available theory . What is a big crunch exactly? well imagine the opposite of the big bang and that is what big crunch is. To elucidate imagine it as a moment in the very far future where the universe reverses its growth and so starts contracting into again a singularity! This singularity is but a different something than what the big bang singularity is if we take into account the entropy value in both the events. Bigbang has the lowest while big crunch has the highest entropy to evaluate. To memorize, black holes are the beast which has not limited their menu only to mass energy but also to the spacetime or fabric of reality itself. Bigbang singularity spewed away all the spacetime (which is still in the expansion phase), this is contrary to what a black hole does! as it gets all the spacetime into it in a manner how water gets into the drainage. To consider the big crunch singularity, it has very much similarity to a black hole.

The dark energy which accounts for cosmic expansion always has a tug of war with gravity as gravity has collapsing effect and expansion the expanding one. Let's say very far in the future (when one and only one black hole remains in this universe and nothing else than some traces of photons) if dark energy weakens and the spacetime curvature or gravity by an ultra-ultra massive black hole (mentioned above) wins the tug of war, then our universe will reverse its expansion to contraction following by the ultimate gravitational collapse into what we call as the big crunch.

Is Bigcrunch the end of humanity then? No! We, humans, seem to be a beast and we will never leave any stone unturned as I mentioned. What we will do then as the universe itself is contracting and ending itself. Here we go, if before the ultimate collapse of our universe, say we could create a wormhole between ours and some other young universe and transfer the humanity over there then could be no end to humanity at least until what I know. Still, to tunnel to some other universe we need to have there all fundamental constants exactly the same as what we have now in our universe, humans during that time would have enough technology to do that still! So the end of the universe may not correspond to the end of humanity or it may be! (provided very far future humans fail to develop that level of technology still). Also, the big crunch is not the only possible fate of the universe! as we have a threat due to vacuum decay, big rip, the heat death of the universe and many more. So we lack precise information about the fate of humanity and our universe or spacetime!

Here in this chapter, throughout the futurism which I have done, I have not violated a single fundamental law of our universe.

The Universe

What is the Big Bang?

Our universe or space-time whatever you say, where did it have its origin from? What was the starting point at which all the forces were united with each other creating a single super force? That's the Bigbang! Every constituent of this universe has their world line intersected at the big bang. There was nothing called spacetime before the bigbang. Spacetime itself got evolved after the big bang took place One may find on youtube the videos of Bigbang expansion being similar to a detonating bomb , but that is misleading. Bigbang took place somewhen 13.8 billion years ago, at that moment how will our universe looked like? It would be like the singularity inside the black hole, the singularity of the big bang is called the cosmological singularity. When I say no spacetime, that means no spatial and time dimension, so here is where our universe was a 0-dimensional entity or say a point. Is 0-dimension the same as the dot made by pen on the paper? No! as if one zooms into it one may get a two-dimensional shape and in fact, zooming more into it will make say some three-dimensional structure visible. We human mind can't really imagine more or less than 3 dimensions and of course, as a part of four-dimensional beings, we can feel time. Thus asking how small is singularity is like asking how small is any infinitely small thing is. Mathematics is the language of this universe, nothing remains unexplainable by mathematics in fact if one gets infinity as the solution then one still technically has a solution which is infinity and that does not mean that maths is wrong! Thus one should imagine the universe at the time of the big bang as a dot on paper made by a pen but still not misconceptualizing it by the same idea . Remember zooming zero dimension doesn’t make it 2D as an example given above .

Big Bang was not a bomb!

Our universe did not start with a blast as may have shown in some youtube videos. At the moment spacetime got into existence, it started expanding at an enormous scale (Bigbang) after which what we call as the cosmic inflation violently but gradually took the work of expansion . In the expansion of the universe space itself is rescaling itself. A balloon analogy of the expanding universe thus would be wrong. The analogy of an expanding balloon with the constant creation of balloon matter in it so that expansion would cause no local stresses in the balloon is somewhat acceptable. Does the expansion of the universe have its effect on the spacetime metric designed by say any planet due to its mass? No! as I said the universe is rescaling itself in the expansion without affecting local curvature or metric.

What makes the inflation period different than the big bang?

The bigbang is the phase when our universe started expanding from an infinitely small point to something at the stage when inflation took in . Inflation is instant in past when the expansion rate of the universe dramatically increased so that expansion is at the exponential scale .

The image below is the three dimensional illustration of the four dimensional BigBang.

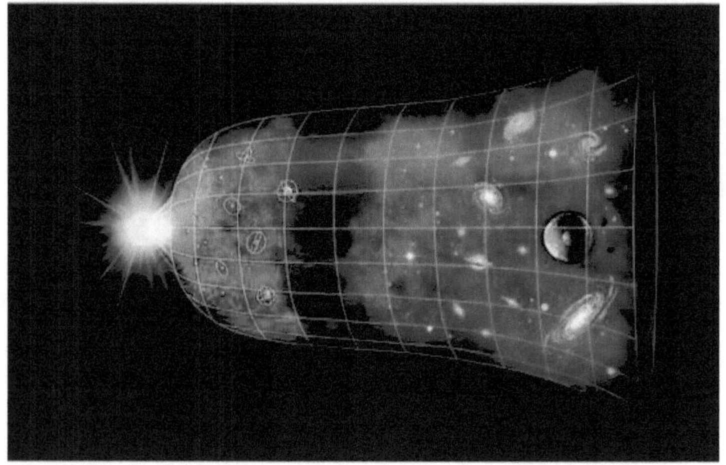

The Cosmic Expansion.

Imagine our universe as a 4-dimensional paper and imagine this paper as getting stretched without getting into it any stress or strain.. Let's start with the question :

How Large Is Our Observable Universe?

The size of our observable universe is 93 billion light years!

But Wait! the current age of the universe is just 13.8 billion years and thus if the speed of light is the highest speed anything can achieve, then how come the cosmic inflation made its way to 93 billion light years within 13.8 billion years ? Well, the answer is that our universe is expanding at a speed faster than the speed of light.

Is that a violation of physics? No! To explain, in principle it is that the inhabitants of the universe including you, me, and every particle in the universe are constrained under the fundamental laws of the universe and not necessarily the universe itself ! Thus the faster than the light expansion of the universe is not the violation of physics.

The History

Einstein after he formulated his field equations, analyzed the meaning of it which suggested the gravitational collapse of this universe. At that time universe was significantly assumed to be something that remains static. Reflecting on the idea of a static universe, Einstein added a term called the cosmological constant into his equations so that the net gravitational pull on the structure of the universe gets neutralized by the repulsive pull of the cosmological constant. Soon (early 1920) after the formulation of field equations, after the construction of many powerful telescopes the movement of galaxies away from the Milkyway was seen. Later in 1929, Hubble discovered that the farther the galaxies are from each other, the faster they move away from each other. After which it was concluded that our universe is not static at all. After which Einstein removed the cosmological constant from his equation. He also called the insertion of the Cosmological constant his biggest blunder. In short, if our universe is expanding and not static, then there is absolutely no need to insert any constant this is what he realized at that time.

How Did Hubble Get To Know That The Galaxies Were Moving Away From Ours?

Even after being viewed by the telescope, the farthest galaxies from us appear as nothing more than a dot in the dark sky then how do we know that they moved past away from us? By the redshifting effect which occurred on the light ray.

Elucidation: Light is basically an electromagnetic field and is connected to spacetime itself as its basic to every constituent of this universe . And if spacetime itself is stretching so does the same happens with the light coming from the galaxy very far away. At local scales the effects of spacetime curvature are negligible but at the astronomical level, it may get changed in the observation. The light from the very distant galaxy that tends to reach us suffers this redshift of light due to the expansion of the universe from which one can calculate the speed at which the galaxies are receding away from each other.

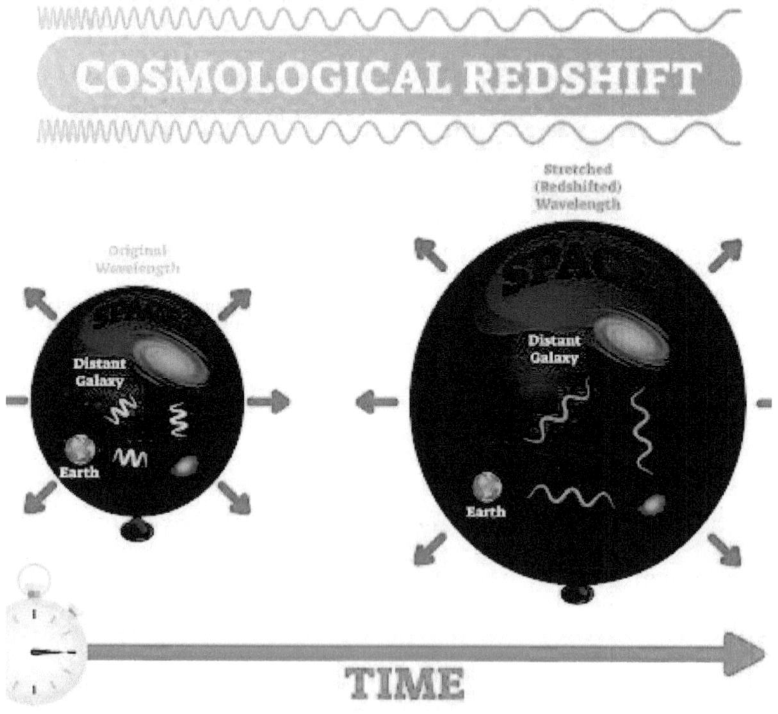

The above diagram may give a rough idea of how the light gets redshifted due to the expansion of the universe.

The constant called the Hubble constant relates to the observable (observed by us) speed of the galaxy or any other astronomical object or event which is receding away from us, which is in fact a function of the distance between the galaxy and we observer. This relation is given by the Hubble law :

$v = H_0 D$

here,

v is the recessional velocity

H_0 is the Hubble's constant

D is the proper distance (between us and the interstellar object or event we are observing)

The conclusion (As per the above equation) is that the farther the galaxy is, the faster it recede from us (due to cosmic expansion).

The Observable Universe.

A Notion Of A Term Distance or length

*Length is basically a physical entity which we all know what exactly is (roughly speaking), still to be clearer it's an interval in space.

How long can a given length span up to?

*Try to imagine this! We Humans cant imagine such things, to be honest. For a length, it can be infinitely large or small. Setting boundaries for how large can a length be is equivalent to predicting how large our universe as a whole is.

What we can define is the observable universe which is well defined for us to imagine and mathematics to play its role. The length of the observable universe is basically our reach of eyesight to the farthest point in the universe at any given time. Our universe is 13.8 billion years old, thus the oldest light we can perceive is always to be the same as the age of our universe. How big is the observable universe? It is 93 billion light years across! The exaggerated (so that one can have an idea) image of our observable universe:

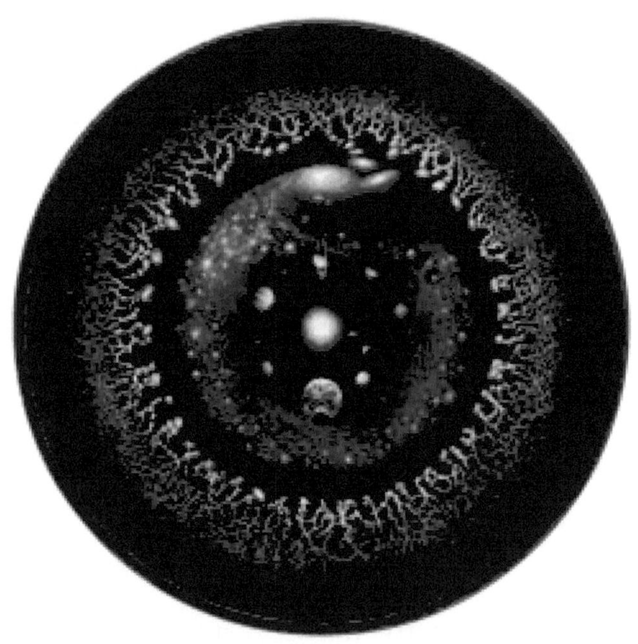

Our observable universe. The circular boundary is the cosmological horizon.

Wait! if the age of the universe is 13.8 billion light years old then the farthest object we could be able to see or the length of the observable universe should be just 13.8 billion light years and not 93 billion light years away as that's the longest length that light can trace since 13.8 billion years and thus how can our observable universe be 93 billion light years across in length? To answer, our universe according to Hubble's law our universe is expanding faster than the speed of light thus a milestone at 13.8 billion years has moved to 93 billion light years away. The light from these moving objects is still able to reach us as they get Redshifted or stretched and are not covering the extra distance stretched due to the cosmic expansion. More clearly (in this case) you can say that the light from 93 billion light years away reaches us within 13.8 billion years! Thus the Boundary of the observable universe which is also called the cosmological horizon lies somewhere 93 billion lightyears away from us. If the universe is entirely taken as a matter of consideration then it can be infinitely large.

What is beyond the absolute boundary of our universe?

Like everything around us say a smartphone, a sofa, or whatever else, we believe that they have some definite size and shape. And it is not only in these cases that we apply the same view but also while imagining our universe. To review, our universe is basically the combination of two different entities space and time. Our universe is thus alternatively can be viewed as a spacetime manifold having some structure. Still, it's not as simple as it sounds, as the universe of spacetime is a four-dimensional entity. Humans can only think of three dimensions and not more or less than that. Applying our view of imagining things that are three-dimensional as having some definiteness simply doesn't apply when we apply it to our universe (as it's four-dimensional). Reaching the boundary of the universe by taking space and time view separately and then total as a matter of consideration:

The Spatial View

Many think there is nothing beyond the absolute edge of our universe. But there is nothing called nothing in science or to say nothing is still something to consider and thus its always something! Many also imagine a kind of a spatial wall to be the boundary but still what if we break that wall by imagination then still what is beyond that is a question. Again some believe that the universe is spatially infinite. We measure distance in terms of numbers which can go till infinity. Infinite distance is unimaginable. It's simply unimaginable.

The Temporal view

Time to consider as one dimensional and dynamic physical quantity we have the same problem. Even if time is considered as being originated from the big bang it can still be wrong as that can just be the parametrization instead of explaining the very existence of time to be originating at the point of the big bang. So it's still a problem that is unimaginable or sensible to human minds.

The Spacetime view

The solution of string theory which says that our universe is just a four-dimensional manifold in the 10 dimensions may answer this. Humans are unable to perceive more than four dimensions. Thus even after if we placed say in 12 to dimension or even more than that then still 4 dimensions is just all that we can perceive. So the answer to the question that what lies beyond the absolute edge of our universe is simply 'The extra dimensions'. These are just the solutions (which are not completely proven yet) of string theory as the existence of the universe require the existence of extra dimensions and as our mental capability what we are seeing is just piece of cake (4 dimensions) of multidimensional hyperspace .

Was the size of the universe well defined or imaginable during the initial fireball (Bigbang singularity)?

Still No! as singularity is supposed to be infinitely small in size so again we are coping ourselves with the infinity which is not really a sensible solution. So at these scales according to quantum mechanics what exists is the quantum foam which is still extra-dimensional in nature.

"Well! we the inhabitants of the universe reside in something and we are searching for what's beyond that something which many say is nothing but still that nothing must still be something to consider as there is nothing as nothing in this universe "

What was before the Bigbang?

This question has several answers to answer. The conformal cyclic cosmology is perhaps my favourite one.

"We humans imagine the entire universe by the view we are always familiar with" and it's the mathematics and our observations that make us agree on the bizarre nature of this universe. The initial fireball (Bigbang singularity) was the point from which the fabric of spacetime was born. If time originated at the time of the big bang then the main question in that sense has no meaning (check the title of the topic!). In our local frames, we always experience a constant ticking of time 1s/s and this is what applies to every inhabitant of this universe. At the scale of singularities, quantum foam is what has the domination and our universe seems to be nothing but one of the fluctuations occurring at that scale.

The Conformal Cyclic Cosmology:

Black holes and the big bang both have singularities with somewhat opposite roles. Black holes swallow spacetime, while big bang singularity is where spacetime had its origin from or was spewed away from. It's being hypothesized that at some point in eternity the last black hole of this universe will spew all the matter which was ever gulped by its entire history. And this is what the big bang happens to be. Very far away in the future when the cosmic recession has begun and when almost everything in the universe collapses itself along with some spacetime if not entirely into an ultra massive black hole (last black hole of this universe). The continuous cosmic recession along with the continuous swallowing of spacetime by the last ultra massive black hole of this universe makes the singularity of this black hole to be a favourable point for the entire fabric of spacetime to fit into and the last black holes singularity of this universe will become a big crunch singularity. And after that Big Bang again ! and then repeat!

During the big bang, the universe had to choose a particular point in hyperspace from where everything began and in the case of the big crunch, the universe tends to collapse at a particular point which is nothing but the singularity of the ultramassive black hole in the eternity. This theory of a big bang followed by the big crunch of the previous universe (aeon) is the Conformal Cyclic Cosmology by Roger Penrose considering our universe to be Closed (that every universe collapses into a big bang). It states that every big crunch is followed by a big bang and so on. The proposed theory originally gives no prejudice to the closed universe and claims that a next big bang is an event that occurs in eternity (And not considering whether our universe is close or open).

Here each period between the big bang and the next big bang (big crunch-bigbang in this article) is called an aeon. And thus each aeon is followed by another and so on in the hyperspace So next time someone asks you what was before the big bang then tell her that it was a big crunch of the universe before ours!

THE CONCLUSION

The first law of Thermodynamics says that the tot). Much of this total energy is in the form of 0-point energy and other is hiding in the bonds (gravity, electromagnetic, weak, and strong) formed by the fundamental system. Currently and many years from now we would be harnessing this energy available to us in form of various systems created by one or many of the four fundamental forces of this universe. Any subatomic particles in this universe can only be identified if it's interacting with similar particles or different via some force. We know that dark matter exists in this universe just because of its interaction with normal matter (gravitationally). Subatomic particles like protons and electrons have a charge as well as a mass which makes them interact via two fundamental forces at the same time (gravitational and electromagnetic force). Protons and nuctrons interact with each other by strong and gravitational force. The neutrino particles interact via weak force which is weaker than E-force and S-force but stronger than the fundamental force of gravity. This makes Neutrinos harder to detect. Even now billions of neutrinos might be penetrating throughout your body and barely any neutrino reacts with mass inside you making its interaction so weak. Thus each and every particle which exists in this universe has an intrinsic property that makes them interact with similar ones and also different. And we humans are currently in a phase of harnessing energy from these particle interactions.

THE SPACE EXPLORATION AND THE BUDGET

We, humans, are continuously striving to make our lives better than what it was yesterday. When it comes to technological development we first give priority to the development of the nation, its security, the welfare of the people, and many more.

In doing so we also explore what is beyond the earth. If any country achieves the status of a developed nation, the focus of the people in the country then remains centralized to other fields like space technology. While building our way to the entire universe we care about a thing called as the budget. The Budget has slowed down our space exploration (Don't get me bad) dramatically. We all know that we cant get the things being freely done from the deepest roots. But, imagine a world where we had nothing called a budget in fields like space exploration, imagine that everything is free in cash when it comes to space exploration. In that world, we could have started the asteroid mining a way before today and obviously also would have terraformed Mars. Today we are somewhere between zero to one on the Kardashev scale. In our imaginary world where budget does not matter, we would have crossed the status of type 1 civilization a way before than what we are today.

The images of the M87 and Sagittarius* black holes and also the images by the James Webb telescope are my favourite ones up to date. The images of the shadows of the black holes were taken by synchronizing seven radio telescopes around the world. Imagine we create the same network by deploying a similar system out in space and with a larger amount of observatories and also a larger distance separating them and thus making the telescope far larger than the size of our planet (as what even horizon telescope did). And now imagine that image of the same two black holes which we would get.

Once we have learned to mine an asteroid, creating structures like a Dyson sphere or anything large enough in space would be not a complication. As we no longer need to extract material for our planet and ship them to our structure. Instead, we have an option to use the gravity tractor to move the entire asteroid to our site and mine the needed elements from it or extract needed materials from the asteroid and move them to our site by the gravity tractor.

We humans don't have a scarcity of ideas but the only thing which hinders our fast progress is our laws which in fact are made for society to work smoothly.

Imagine you are on your trip to the Sirius star, traveling near to the speed of light. After you return back to the earth, all your peers would have died and your grandson is of your age now. That seems so unreal but in fact, that happens in reality. Science and technology are just not all about biology or developing new technologies for making our life smooth but to explore the entire universe at least in terms of imagination and by asking questions. The question is which can bring new theories and build technology that will allow interstellar travel throughout our universe. Currently, we are tracking our way back in time by using the James Webb telescope and aiming for mars. This progress although appreciable but is still very much low than how it would have been when the budget (budget to run human life smoothly) is not hindering our exploration of the entire universe. Humanity must not carelessly leave this space exploration technology to the future civilization. It's possible now! if we strive for it, space exploration is our need and we should not trap ourselves in a dot-sized planet (as compared to the astronomical scales).

Some Interesting Stuff

Elementary particle physics (Introduction).

Having an atom into existence is not just the game of an electron, proton, and neutron. Proton and neutrons are made of 3 quarks, to hold or glue this quark one needs a gluon particle (strong force carrier). For an electron to maintain its energy level or say to interact with the proton in the atom itself needs virtual photons (electromagnetic force carrier), when an electron combines with a proton a neutron and a neutrino particle is produced. Again these particles have their own anti-particles. To give the fundamental particles like electrons, quarks, etc their masses we have Higgs bosons, which are also called a God Particle. These are just a few of the large sets of particles in existence. In reality, there can be innumerable particles in this universe, some of them are very mystical like the dark matter as we don't what exactly are made up of. The standard model of particle physics is all about it. How these particles interact with each other , etc. According to QFT (quantum field theory), the fundamental particles are nothing else but excitation in the quantum field. The Higgs field is a specific type of quantum field which gives the fundamental particles, their masses. The discovery of the Higgs boson was one of the greatest achievements of all time. Physical detection of such particles crystalizes the prediction made by the mathematics.

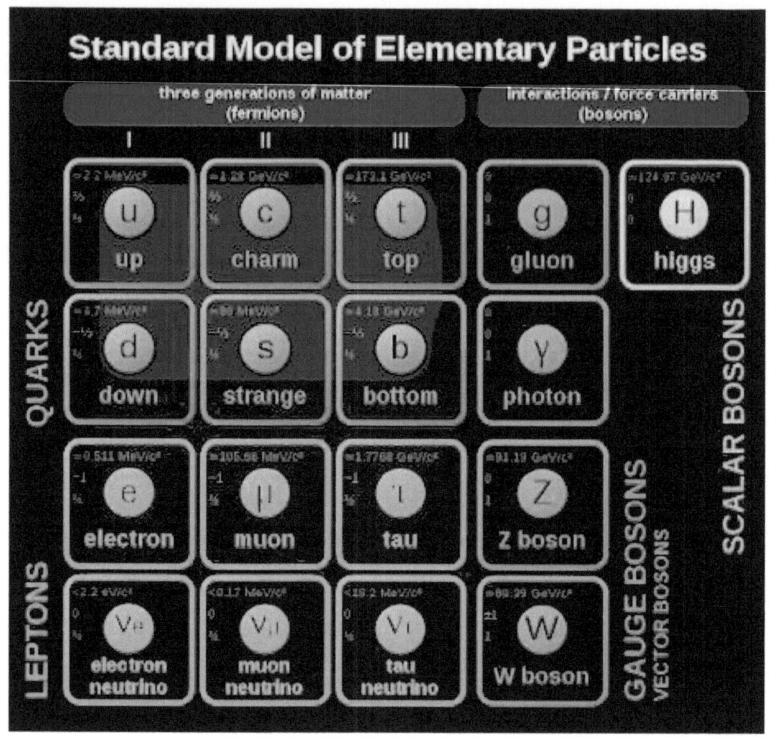

The standard model of particle physics

Just after Bigbang, the Higgs field was inactive and thus every particle in that epoch were massless like photons and gravitons . Having no mass every particle there travelled at *c* (the speed of light). The large Hydron collider (LHC) is the particle accelerator which is located in Geneva, Switzerland. In LHC two or more particles are collided to get a stream of new particles being released at the instant of collision. One of the best experiment detection which makes up the holy grail of physics is the experimental discovery of graviton.

Feynman Diagram by Richard Feynman which discusses various particle interactions in form of particle interaction diagrams as given below

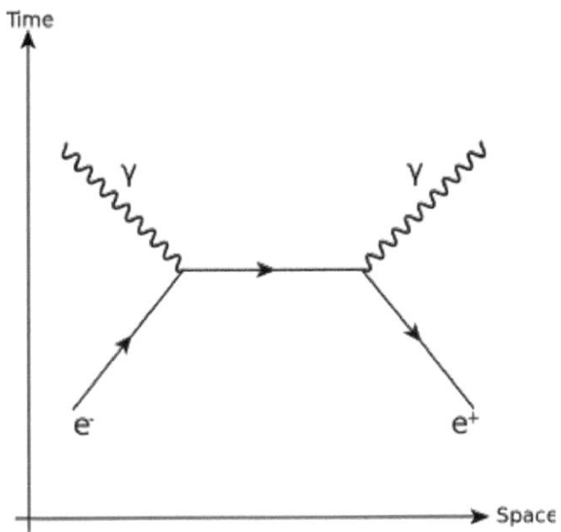

Here in this diagram, an electron pointing its temporal trajectory in the future interacts with a photon and instantaneously turns itself into a positron . If the current in one direction

due to the negative charge of an electron is taken as positive can be considered as having a negative value if we had a positron instead of an electron as the charge is flipped (mathematically one can check). Thus when an electron flow through spacetime gets deflected if changes its direction then can logically be considered as the flow of positron instead. That's so interesting! an electron traveling backward in time is a positron ! To consider any particle there can be infinite Feynman diagrams associated with it. Including the standard model, QFT, and Feymann diagram there are many theories and mathematical ways to describe the behaviour of subatomic and elementary particles. To conclude, particle physics gives us a lot of information about the time just after the big bang, also it's proving a great way to formulate a unified theory and many more.

What are the Light Cones?

Let's take two axes, one is space, and the other is time. The slope of the graph on this plane is what we call the velocity. At average speeds, we get some well-defined curves as our graphs, but for the speed say at the speed of light the velocity graph become parallel to the time coordinate (one can check). Now rescale the same coordinate system such that the graph of the speed of light is a constant graph aligned at 45 degrees concerning both the axis and not something parallel to the time axis. Now we have some new coordinate systems which are also called the Penrose diagram . To deal with light, in this coordinate system light ray always travels at 45 degrees. In this universe, nothing can travel faster than light even the information. Thus the fastest way we can receive information is by the light itself. For example, if the famous Betelgeuse Star ends its life now (Let's say), we still are not able to know that until that information is reached to us by the light, so if one is has a view on Jupiter then basically she is seeing the history of Jupiter and not what exactly is going on there say a second or less than that ago. Thus to conclude, events are relative.

What are the lightcones?

If we analyse the motion of an astronaut in the Penrose coordinate system then that particular graph is basically called as the world line of the astronaut (Generally speaking). As I said, for any information to reach the astronaut by the fastest way then it has to be communicated via a ray of light. In our new coordinate system light always travel at 45 degrees, so here we conclude that for any information to be communicated to the astronaut, it has some speed range to covey the same. This speed range is less than or equal to the speed of light. Equivalently, in terms of the Penrose diagram it has to be in the range of less than or equal to 45 degrees it's the path of light it travels, and beyond that is not possible in this universe. This range of fastest or slowest communication in terms of Penrose diagrams forms two cones with slant height inclined at 45 degrees. Here, one come remains above the observer's present spacetime location and other below it. This is is what a light cone looks like:

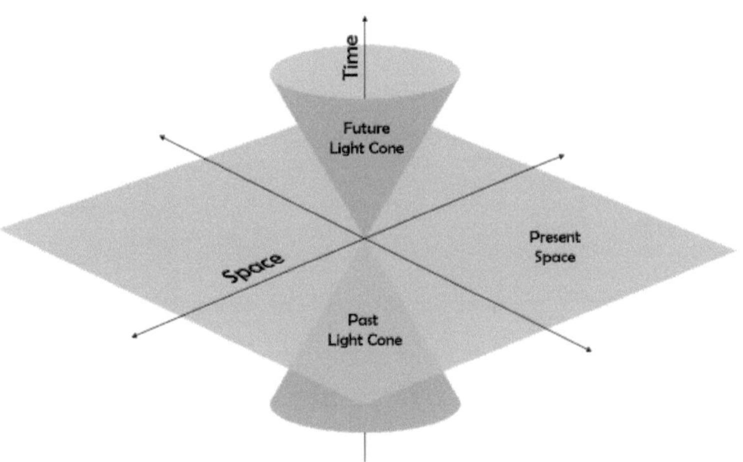

Thus for an event to communicate to the observer, it has to be in her lightcone otherwise the information remains uncommunicated.

Above, we considered an observer who is at rest in space with ticking time. Now we consider both, changing space as well as time. According to the principle of relativity, laws of physics remain constant in any given frame and so does the lightcone for that particular observer. Thus if the astronaut changes his spacetime components, he carries his lightcone along with himself by tilting it along his worldline .

A notion of what distance exactly is.

This article will change one's view of the commonly well-known quantities in this universe. Reader's notion of a physical quantity 'Distance' will dramatically change after reading this .

Not many know how the normal life term distance seems to be at the astronomical and tiniest scales of space. We generally perceive distance as an interval in space. And we represent this quantity by the numbers. Numbers can go from absolute zero to infinity both of which seem to have no meaning in terms of mathematics. In that sense when we say that distance is 0 then what does it mean? is it 0.0000000000000000000000000000001 km (say) but that's still not an absolute zero? Absolutely zero means infinitely small and as we can't imagine what infinity is so does absolute zero.

There remains no reason to how long or small (like numbers) any distance can be. When mathematical terms like infinity or absolute zero come into something which seems to be so well established or at least imaginable at local then it seems bizarre and makes us think what the heck is out there!

The Minimum possible distance

According to quantum mechanics, Planck length is the shortest length possible. This is the solution of equations of quantum mechanics for physics to be well-established over the places like the cosmological and black hole singularity. The value of Planck length is 1.616 255 x 10-35 m. Well! you imagine the thickness of human hair as the size of the observable universe and then the Planck length would be 0.1 millimeters. It's incredibly small! But still, it's not that easy. Suppose you shrink yourself at the Planck scale then you find the Planck length as the length which is indivisible. That is so bizarre! We believe that every length in this universe can be divided into any number of segments but not so in the case of the Planck length. Think and try to imagine the Planck length, if you could do then you are a superhuman!

Our universe is expanding so it's still scaling wider and wider. And we don't have any predicted limits for the maximum possible length. And amusingly, even if we have then we humans will ask why can't something be larger than this length.

“Nothing in this universe can be at the absolute rest.”

To talk about a position say of any object which is at the rest which we also represent mathematically as v=0 is not really at the rest. Let's take a solid wall around for example and the question is whether it is at rest? No! as the wall is made out of individual atoms which at the tiniest scales cant be really at the absolute rest . And what keeps the wall compact is basically the electromagnetic interaction between each and every atom in it which can still be maintained even if the atoms or their sub constituents are not at the absolute rest. Thus ultimately nothing is at rest. Physical quantities cant really be zero in value as zero itself is not zero.

A notion of what mass exactly is.

The physical quantity 'Mass' is only discussed here and others in subsequent articles.

There are so many things in this universe that we know what they are but still, we don't know what they actually are. One of these things (Mass) will be discussed here.

Mass

It's nuts to say that mass is weight. Every object in this universe which is made of matter or elementary particles has a measure called mass. Many say that mass is a measure of the amount of matter contained in the object, which is misleading! Mass is not what many think it to be. When you touch some object you say that you are touching some mass, which is wrong as the feeling of touch is due to electromagnetic interaction between atoms of your finger and the thing you are touching so there must be some gap over there. Thus one in reality is not really touching a mass.Lets ignore the electromagnetic force , and the question is what is one touching now ? An atom ? An Atom is made of basic sub constituents electrons, protons, and neutrons. Protons and neutrons are made of a combination of three quarks. That's it and we don't what the subconstituents of quarks and electrons are (excluding the solution of string theory). But still, these indivisible particles possess mass! If the mass is the amount of matter an object possesses then what kind of mass do electrons, quarks, and many indivisible particles posses ? Mass is not the matter content !

The mass has the property of exerting a force (not really a force) on each and every mass which we call gravity. Gravity can alternatively and less clearly be defined as the tendency of two masses to attract each other. The mass can again be defined as the curvature in spacetime, it can be defined as the energy itself. Remember it's the curvature in the fabric of reality! Mathematically we may describe what mass is but a physical picture seems unimaginable. Some particles like photons and graviton even come out to be having no mass. Mass is relativistic, it changes as one accelerates. Mass is a very common term in Newtonian physics but ultimately we don't know how to give it a pictorial form or even imaginable. In quantum mechanics, one can have a particle at two different positions at the same time i.e. the same mass at two different points at the same instant. The ordinary definition of mass in this sense seems nuts. Many of us probably accepted this definition in high school just because we thought we know about mass but that's not the case at all.

What the heck Mass exactly is?

One may have asked the question 'what is mass? but may have not got the proper anwer to the exact answer to the question addressed .

To begin with, I would first introduce the reader to the term 'Energy' for a better understanding and experience

Energy is basically the ability of a system to get some work done. There are four fundamental forces in this universe:

1) Electromagnetic force

2) Gravitational force

3) Weak force

4) strong force

Force is mass times acceleration and energy is force times displacement. From this , we conclude that energy is associated with some distance or movement of the particles under the mentioned fundamental forces. Digging into the topic still further, The First Law of Thermodynamics says that

'Energy can neither be created nor destroyed'

or alternatively

'The total energy content in this universe remains constant.'

So the amount of energy in this universe will never vary. And what we are seeing is just the forms of the same energy. Energy is not what one can see and it is just the ability of the system to perform some work.

Energy is also associated with an object which has some velocity this energy is called Kinetic energy. Is kinetic energy a relative concept? yes! we think that we all are resting in just our frame, but from the distance very far away from the earth, we are moving at a tremendous speed. So it can be said that from that frame of reference we have some kinetic energy and from ours, we have no kinetic energy

Special relativity :

Special relativity says that energy is the same as mass and vice versa!

$$E=MC^2$$

Thus so much energy is associated with just a small mass. Imagine this tremendous amount of energy released by some small amount of mass being converted into the other forms of energy like heat, kinetic, etc and we have so much work done!

What is mass?

The equation $E=MC^2$ is mathematical and rigid! (cant derive here). The equation explains rest mass in terms of energy (the ability of the system to do some work) which is something perceivable (at least)! One should wait think on this to get again a clearer picture. To express, $E = Mc^2$ describes the rest mass in so effective way. One can get a still clearer picture if one checks the derivation (one can skip the derivation!). To summarize, the derivation starts with the fact that mass is a relative concept from which the total energy equation is given by the particle's rest mass energy added with the kinetic energy it has. And as kinetic energy is zero when the body is at rest the total energy remains equal to the rest mass energy of the particle giving us the mindblowing E=

Mc^2. Einstein's mass-energy equation defines rest mass in terms of energy which is the ability of the system to do some work , that is so great! To conclude, mass is the energy and the energy is to be more precise the ability of the system to do some work. Through out the article I introduced reader with what energy is and by the way of mathematics I derived mass energy equivalence equation (verbally) after which I given some physical picture of the whole thing

Why are the galaxies like ours (Milky way) flat, while the planets and stars spherical in shape?

If you are in hurry then just read the summarization mentioned down and that's it (sufficient to answer the main topic of this article) Although reading the main content has its own taste and deep understanding. There are galaxies with a variety of shapes in our universe. This article is limited to only spiral galaxies (eg our Milkyway galaxy).

The laws of physics are the same for each and every constituent or inhabitant of this universe. Dating back to several billion years in the past (even before when galaxies were forming in our universe). That was the time when individual particles were interacting with each other gravitationally to form some cloud-like structure which is our future galaxy and still further processing will get our planets to us. But what mechanism gives our galaxy the disk structure it has (the spiral galaxy), while most of its inhabitants like planets, stars, and also some asteroids are spherical in shape.

The underlining force which clumps the matter into galaxies or planets is basically the same (gravity). Then why is there a difference in structure? In a nutshell, it's just due to the way and condition of formation.

The Formation Of Galaxies :

The formation of the spiral galaxies is a much slower process as compared to the planets and stars. In the case of galaxies, individual particles (separated by wide distances) interact with each other gravitationally, until it forms a rotating supermassive black hole. Once the rotating supermassive black hole forms, the chain continues further. The centrifugal force due to the revolution of particles around the axis of rotation of the SMBH (supermassive black hole) maintains the matter in the vicinity uninteracted. This structure of revolution of particles around the rotating SMBH further gets a disk-like appearance. The influence of matter above the disk on what is inside the disk nulls off due to the interaction of matter which is below the disk (equal and opposite forces). The only net forces now remain is the gravitational influence of the matter inside the disk on the matter above and beneath the plane (downwards). Thus the net force on the matter below and above the disk remains downward towards the planet of revolution. This overall process gives a system a disk-like look. Refer to the following image:

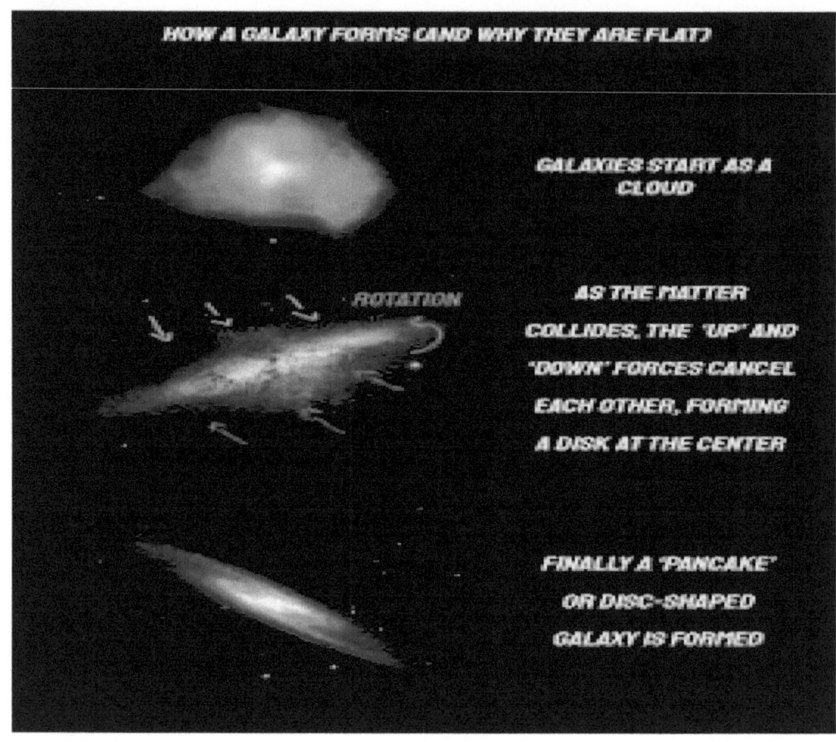

The Formation Of The Planets :

In the case of planet formation mostly what collides are the bulk particles and rarely finer ones. Again as the matter collides it gains some angular momentum to itself and individual bulk particles say above and below the plane of rotation approach the system and combine with it. In the case of the formation of galaxies the particles were finer and were assumed by us to be symmetrically oriented although widely separated (a very close situation to what is in reality) which makes our force cancellation statement a foundation to result. In the case of planets, the particles are bulkier. Thus the probability of symmetric orientation among particles is pretty less, so the force doesn't cancel out over here and and the bulkier particles combine to form a spherical planet .

The Formation Of The Stars :

In the case of a star, the matter interacting although is small in size (maybe of what is in the case of galaxies), the difference that appears is the formation of SMBH . In case of star formation SMBH do not form . Thus no accretion disk . And what remains is just symmetric attraction of particles equal from all side , which gives star a structure of a perfect sphere. And once the fusion reaction starts we have our prototype star.

The Summarization :

In the case of planet formation, bulkier particles are unsymmetrically distributed and thus there is a net inward force on every chunk out there. And as the gravity is the same from all sides the plane takes a spherical shape.

In the case of stars, the particles are finer and less widely separated and also are symmetrically distributed. And as there is no formation of the violent disk due to absence of SMBH . All matter is attracted symmetrically

In the case of galaxies, the atoms are finer and very widely separated from each other. Also, the particles during formation are symmetric only at large scales. Any rare event of finding a concentration of these particles at a particular point (say) makes that point the center of the galaxy. As time passes due to the constant clumping of matter, a rotating supermassive black hole forms which result in the formation of an accretion disk (disk in which highly dense matter revolve around the black hole) around it. The force of gravity due to matter above and beneath the accretion disk cancels out due to equal and opposite directions of the force. Thus the net force of the matter above and below the accretion disk is towards the disk due to the gravitational influence of the accretion disk . And thus the whole stuff becomes the part of that accretion disk that we call our galaxy. Thus even if the underlining force behind behind the formation of stars, planets and galaxies is the same, the nature of the matter , its position in space , distance between individual particles are what decides the shape of the structure formed.

The Multiverse Hypothesis And The Many Worlds Interpretation .

In infinity war, Thanos shifted a huge part of an entire population to some other universe. But does a multiverse (which consists of many universes like and unlike us) exist? The fact that our universe is not alone in the hyperspace (extra dimensions) is the Multiverse Hypothesis. Thus we have upgraded our union set of universes to the multiverse.

The many worlds interpretation:

Quantum mechanics is the language of probability. It assigns every particle,, not a discrete existence but a waveform. This wave is what the particle is and vice versa. The low and high peaks of the wave including together represent existing universes. So our universe is just one of those peaks. It would be wrong to imagine the wavefunction of the multiverse as a 2 or 3-dimensional graph, as one should remember that we are talking about the wavefunction in Hyperspace! , The reality which we perceive is due to our mere observation which collapses the wave function to produce some discrete results. One might question then that what about the other possibilities ? where do they go even if they were not made discrete by our observation in our universe. The collapse of function is not only dependent on the consciousness and the observation of the observer but also on the environmental conditions around it. Thus the events which had no discreteness in our universe are what are at the peaks of some other universe. So maybe a Schrodinger cat is alive in our universe, but the dead part of it is somewhere in another universe. Thus to comment on the wave function, no part of it is unreal and it's our observation and the conditions around which make the particular part of a wavefunction to peak and achieve discreteness or to collapse in ours, if not in another universe. In short wave function is a combination of the various outcomes in the various universes in the multiverse.

A 5th dimensional perspective

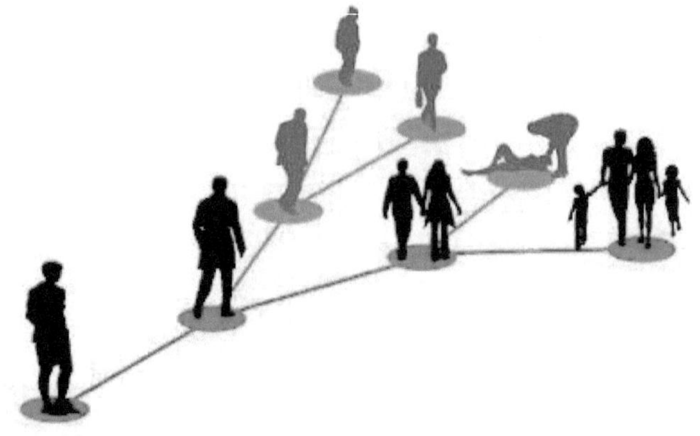

Illustration of diverging timelines

when one goes at the fifth dimension, one can see how our future is defined by various wavefunctions and our observation

What is it would like to live in some other universe in a multiverse?

The fundamental laws of physics including the universal constraints like the speed of causality (3,00,000km/s) and many more which define our universe may not be of the same value in another universe or they might not even exist there. The fundamental laws of these universes might be very different than ours. So don't get shocked if you come to know that the value of pie is 100 in some other universe or say time runs backward there. The energy created in our universe is constrained to follow a set of universal laws which define the bodies of you, me, and everyone and everything. And Thus going to some other universe to stay will simply against our existence, as our body is defined by the set of universal laws, and going to some other universe with a different set of laws will simply deny our existence. But what if you could find a universe with laws of physics including the fundamental constants like ours with very high precision? The answer is still no. However precise you may be in determining the fundamental constants, say the value of a particular constant is the same until the 15 decimal place, but still a slight variance in 16 or say 20 decimal place will still deny our existence. And since this is our inability to measure or compare any constant to the absolute preciseness, it's only our experiments that can prove whether any universe in a vast multiverse is liveable.

The Evidences Of Another Universe

We speculate that by some of our observations like a cold spot in the Cosmic Microwave Background, other universes or multiverse can exist ! . Our observation proves the explanation of a cold spot in the cosmic microwave background (map of our universe) due to some void in space to be wrong. And thus we speculate that the resulting cold spot is some sort of a past impact of our universe with some other.

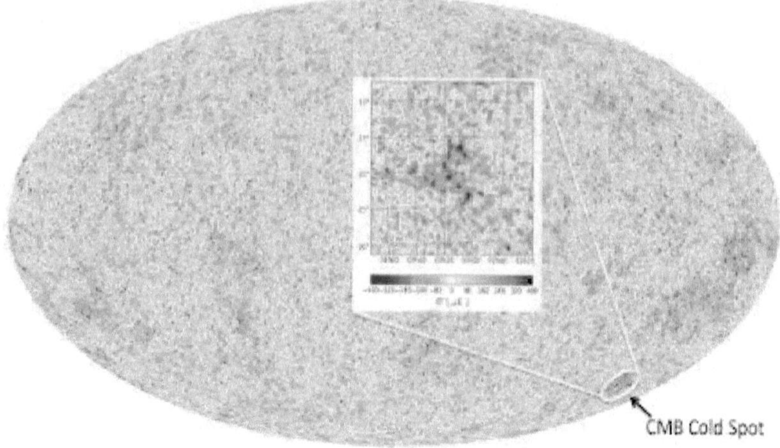

CMB cold spot could be the sign of the possible collision of our universe with some other

The Participatory Universe.

To understand this content, the reader should have prerequisite knowledge of the famous observer effect. And if anyone doesn't then plz visit the chapter 'The Observer effect'

Observation crystalizes the wavefunction of multiple possible realities into what we perceive as a reality. Shortly, reality doesn't exist until it's been observed. And the laws of physics are made in a way that will allow the existence of the observers who in fact had created it.

Where there the observers before the big bang?

Well as I said in my observer effect article that observation in generality is the sense of our existence today which spreads throughout the past of the future of the entire universe. The result that humans or conscious observers are alive today is due to the highly unprobable setup of the universe and the conditions. And somehow our presence today is the observation that has been crystallizing the wavefunction into the setup which allows our existence. So even if earth came somewhen around 4.5 billion years ago, it's the observation of our existence today and the future that the universe existed in the way that it is. This is called the Anthropic Principle.

Get your scales from 4D human beings to 5D beings and look at the time as a set of individual planes separated by some distance. where each plane represents your past, present, and future. One can imagine it as the past, present and future exist simultaneously. Imagine it (As a 5D being) as you're looking at your past from when you were in the kinder garden from what you are now to the time when you are very old. If our universe, in reality, is like this then it is our present existence that arranged the laws of physics and the other conditions in a way that allowed our (Observer's) existence and the same applies to the future beings whose observation of the laws of physics and the universe is what the way today and now it is. The idea that the fate of the universe is dependent on its participant (we observers) is the Participatory Universe Hypothesis.

The representation of the participatory universe

Our universe in reality does not need to be in this shape. This is just the representation that the very point of human existence through his timeline has acted as an observation(represented as an eye in the diagram) which had spread until the very dawn of the universe when the laws of physics were set up to create the universe which is accompanied by the observers like you, me and everyone.

Are we alone in this universe?

To this date, there is no solid evidence (Although many claim there are) of the existence of aliens. But still, are there any one of them out there? Any planet in another system in the habitable zone (zone suitable for the creation of life) gives us the chance of having extraterrestrial life. Our planet earth, which is in the solar system, is in the miky way that has more than 3200 star systems in it and still there are more than three trillion estimated galaxies in the observable universe. Thus in that sense, it's highly probable for the alien lives to exist. Can Extraterrestrial life if exist be more intelligent than us? well! we don't know. Aliens may even be type 2 civilizations at the level of which technology is enough advanced to harness the complete power of any star. Aliens also may be advanced human beings. If they meet us we would have a great opportunity to learn from them. Also, they can be bacteria or any unicellular animal too.

Jupiter's moon Europa is quite likely to sustain life in it as there remains the liquid water underneath its ice sheets which have widths of several kilometers wide. Europa gets sunlight but not that bright to maintain the water in a liquid state beneath So from where does Europa get that energy to maintain the liquid water beneath the ice sheets? the answer is the tidal gravity of Jupiter. The tidal gravity or tidal force of Jupiter brings out some deformation in the shape of Europa, this mechanical energy is then converted into heat energy by breaking various molecular bonds providing some heat energy source to maintain the liquid water under the ice sheets of Europa. One also can compare this with the tennis ball when it gets hitted by a racket. When tennis ball is hitted by a tennis racket, it gets deformed and infact get heated due to the mechanical deformation acted on it. In the same way Europa receives it's heat by which it maintains liquid water beneath the icesheets on it. But here what is deforming Europa is the tidal gravity of Jupiter. This deformation is not as great as in tennis ball analogy but quite enough to provide energy to the planet to keep water in the liquid form beneath the ice sheets on it. A mission is planned by NASA till 2024 to study still in detail to investigate the suitable conditions for the existence of life on Europa and if we get some positive results there, then it would truely be one of the greatest achievements of us.

Well! what about the aliens from some other universe? Great idea right! If the multiverse exists, then we have a greater chance of life harboring in each of the individual universes in it . So if we are alone in 'our universe ' then there might be aliens existing in another universe too. What about the bulk beings? (Multidimensional beings) There is no proof of the existence of bulk beings. If God exists in 10 or 11 dimensions we may call him a bulk being or bulk aliens or whatever you like. There is a famous equation of aliens which gives the probability of the existence of extraterrestrial life in our universe. It considers each and every parameter like the number of planets which are habitable, the fraction of stars orbited by planets, etc. This equation is also known as the Drake Equation.

$$N = R_* \cdot f_P \cdot n_e \cdot f_l \cdot f_i \cdot f_c \cdot L$$

N = number of civilizations with which humans could communicate
R_* = mean rate of star formation
f_P = fraction of stars that have planets
n_e = mean number of planets that could support life per star with planets
f_l = fraction of life-supporting planets that develop life
f_i = fraction of planets with life where life develops intelligence
f_c = fraction of intelligent civilizations that develop communication
L = mean length of time that civilizations can communicate

The presence of us in this universe as the only life in existence in actuality is a rare case. If we are in reality all alone in our universe, could it be interpreted as there being some relation between the existence of humanity and the existence of the universe? Roughly speaking it could be! Even some observations of cosmic microwave background like the axis of evil suggest us that we humans have some special role in this universe! But still, this is speculation and there is no solid basis proving it strongly. But until we investigate each and every suspected habitable zone for life, our hunger of searching for the extraterrestrial life will persist.

www.ingramcontent.com/pod-product-compliance
Ingram Content Group UK Ltd.
Pitfield, Milton Keynes, MK11 3LW, UK
UKHW041839190726
13854UKWH00002B/621

9 798888 837931